PROFESSIONAL ENGLISH OF Building Environment and Energy Application Engineering

建筑环境与能源应用工程专业英语

张 琨 / 编著

大连理工大学出版社
Dalian University of Technology Press

图书在版编目（CIP）数据

建筑环境与能源应用工程专业英语 / 张琨编著 . --
大连：大连理工大学出版社，2021.1
ISBN 978-7-5685-2647-0

Ⅰ. ①建… Ⅱ. ①张… Ⅲ. ①建筑工程—环境管理—
英语—高等学校—教材 Ⅳ. ① TU-023

中国版本图书馆 CIP 数据核字 (2020) 第 150728 号

大连理工大学出版社出版
地址：大连市软件园路 80 号　邮政编码：116023
发行：0411-84708842　邮购：0411-84708943　传真：0411-84701466
E-mail：dutp@dutp.cn　URL：http：//dutp.dlut.edu.cn
大连图腾彩色印刷有限公司印刷　大连理工大学出版社发行

幅面尺寸：170mm × 240mm　印张：15　字数：140 千字
2021 年 1 月第 1 版　2021 年 1 月第 1 次印刷

责任编辑：王　伟　责任校对：婕　琳
封面设计：冀贵收

ISBN 978-7-5685-2647-0　定　价：38.00 元

本书由大连市人民政府资助出版

The published book is sponsored by the Dalian Municipal Government

Introduction
前　言

建筑环境与能源应用工程专业原名为供热、供燃气、通风及空调工程，后改为建筑环境与设备工程；2012 年改为现名以后，其外延又得到了极大的丰富和拓展，主要研究建筑的供热通风空调系统、建筑设备智能控制系统的设计原理与方法、施工安装与运行管理方面的基本知识，以及节能分析，使建筑内部环境健康、绿色、节能。中国建筑节能协会能耗统计专业委员会发布的《中国建筑能耗研究报告（2019）》指出，2017 年，我国建筑能耗 9.47 亿吨标准煤，占全国能源消费的 21.11%；建筑碳排放 20.44 亿吨 CO_2，占全国能源碳排放的 19.5%。为了降低能耗和碳排放，保护绿水青山，建筑环境与能源应用工程专业的重点日渐与土木建筑和自动化控制等专业交叉融合，趋向建筑节能及环境保护和可持续发展，实现最高效率地利用资源，最低限度地影响地球环境。

本书作为建筑环境与能源应用工程及相关专业的英语教材，融合了经典理论、工程技术、科技前沿、实际应用等多方面知识，具有以下特点：

（1）按照由浅入深、循序渐进的原则，既继承了传统图书的经典思想和理论（熵增原理、卡诺循环、傅立叶定律、伯努利方程等），又融入了本学科最新前沿知识（热泵原理、相变储能、辐射换热等）。

（2）在每一章中都提供了大量、丰富、细致的系统、产品和局部的图解并加以说明，便于引导读者身临其境地感受现场环境，有助于降低学习难度，激发学习兴趣，以增进读者对理论知识和工程实际的理解和掌握。

（3）总结归纳了行业最大的学术组织——美国暖通空调工程师协会的相关信息，便于读者紧跟时代，有的放矢。

本书对于建筑环境与能源应用工程、能源与动力、热能工程、制冷等专业人员掌握英语词汇、增长专业知识、拓宽国际视野具有积极的作用。

全书共分 14 章。引言概括性地介绍了能源的科学问题，引发读者思考。第 1~3 章介绍热力学、传热学、流体力学的专业基础知识。第 4~7 章介绍制冷、地板采暖、通风、空气调节等技术特点及其工程应用。第 8~13 章介绍热泵、冷冻水、相变材料、热舒适性、辐射换热等最新技术的发展与应用。第 14 章介绍美国暖通空调工程师协会的成立、发展及未来趋势。

特别感谢教研室同仁张殿光、田兴旺、李丛、孙丹等老师对书稿内容的支持及高兴老师在工程实际方面对本书提出的中肯意见和建议。感谢大连市人民政府和国家自然科学基金委、国家留学基金委的资助。感谢家人的长期支持。

由于编著者学识有限，书中难免存在不妥之处，敬请读者批评指正。

编著者

2020 年 6 月

Contents
目　录

Introduction:
The Science of Energy
导言：能源科学

The energy of a system is, therefore, sometimes briefly denoted as the faculty to produce external effects.

—— Max Planck

Since the industrial revolution two hundred years ago, the overarching achievement of the world is surely the substitution of machine-assisted energy for human and animal drudgery. Machine energy dominates modern agriculture and factory work as well as everyday city living. Plowing, sowing and harvesting machines have taken the place of humans, oxen and horses; washing machines, dishwashers, automatic dryers and vacuum cleaners have replaced both class-based maid-service and dreary housework; and automation has relieved and eased much backbreaking factory labor.

How did we come to understand energy sufficiently to make such a **fundamental**

fundamental
基本原理

development possible, and what is the source of this energy? There are, of course, many economic and political issues having to do with access to energy (or the lack of it), such as the short-term power and long-term deleterious social effects that enormous oil reserves may have on societies with few other sources of income, and the enormous effects of differential access to sufficient sources of energy on the power relationships between countries. In addition, there are important engineering, social, and political problems such as the dangers of nuclear fission reactors, the disposal of radioactive waste produced by them, and the environmental damage caused by surface coal mining.

But to be clear about the basis of all these questions it is surely good to understand the scientific fundamentals of energy and that is what this book is about. It will explain both the scientific laws governing the use of energy and the science involved in its utilization, as well as the cosmic history of energy. Written for readers with little scientific knowledge or background, it deals with the scientific aspects of energy, such as its definition and underlying laws as well as its various forms, its storage, and its transport. We will be discussing matters of physics and chemistry, as well as some biology.

Thermodynamics elucidates the origin of the concepts of work and other mechanical energy in Newton's mechanics, the two important laws of thermodynamics, conservation of energy and its degradation for the purpose of actual use, as well as Einstein's enlargement of what constitutes energy by his famous formula $E=mc^2$.

When we are done, the reader will understand both the basic scientific ramification of energy and the fact that it is ultimately the Sun to which we owe almost all the energy we use on the Earth, one of the few exceptions being nuclear energy, the very source that also fuels the Sun.

本章重点内容介绍

两百年前的工业革命，使得机器辅助的能源代替了人类的繁重劳动。

本书将解释关于能源使用的科学规律和利用能源所涉及的科学，以及能源的宇宙史。它涉及能源的科学方面，如能源的定义和基本规律，以及能源的各种形式、储存和运输。我们将讨论物理和化学以及一些生物学问题。

热力学阐明了牛顿力学中功和其他机械能概念的起源、热力学的两个重要定律，以实际使用为目的的能量守恒及其转化，以及爱因斯坦著名的公式$E=mc^2$，拓展了能量的范畴。

Chapter 1
Thermodynamics
热力学

1.1 Introduction
简 介

工程热力学是研究热现象中物质系统在平衡时的性质和建立能量的平衡关系，以及状态发生变化时系统与外界相互作用的学科。

Thermodynamics is a physical science; that is, the **principles** that form the framework of thermodynamics are all based on observations of physical phenomena. Following the observation of a phenomenon, experimental evidence is collected to verify that the observation was indeed a correct one. Finally, once the principle has been accepted, the physical observation can be recast into a mathematical **formulation** that will provide a mechanism by which the principle can be applied to engineering problems.

thermodynamics
热力学

principle
原则

formulation
配方

A large portion of the subject matter of thermodynamics deals with a study of energy and its conversion. In fact, many people define thermodynamics as a study of energy and its relationship with the properties of matter. While most people are familiar with the concept of energy, few are able to give a rigorous definition of energy. On a very simplified level, energy could be defined as a capacity to produce change. The energy output of an **automobile** engine provides the capacity to move from one location to another. The energy output of a power plant provides the capacity to produce a wide variety of changes — to operate motors, television sets, and lights, to name only a few possibilities. Energy derived from petroleum products can be used to power many different devices. Solar energy provides a capacity for change by heating water and air for comfort purposes.

The basic principles that are the starting point for the study of thermodynamics are the conservation of mass, the conservation of energy, and the second law of thermodynamics. The conservation of mass and energy are usually discussed in some detail in **introductory** courses in physics, so most students are somewhat familiar with these basic principles. The second law of thermodynamics, however, is usually unique to a course in thermodynamics, and it is a basic principle that was developed from the physical observation that without external sources of energy, heat transfer always occurs in a preferred direction; that is, heat transfer is always from a region of high temperature to a region of lower temperature. From this observation the concept of entropy can be formulated and used to

automobile
汽车

introductory
引导的

predict whether a particular process can occur and to what extent the process will occur.

工程热力学涉及传热、做功、动能、势能、物性等方面。

Thermodynamics provides important relationships among heat transfer, work interactions, kinetic and potential energy, and quantities that are called properties, which describe the condition of any substance. In fact, a major **contribution** of thermodynamics is the mathematical relationship between the amount of energy that is transferred to a substance and the change in the properties of that substance. This relationship is used to study the operation of devices that utilize and transform the various forms of energy. Thermodynamics is therefore particularly important in an era of dwindling supplies of readily available energy and increased interest in energy conservation.

In the **preceding** paragraph terms such as work interaction, heat transfer, kinetic energy, and potential energy were used without definition, but most students should be familiar with these terms since they are used in physics, statics, and dynamics courses. Rigorous definitions of these terms are included later in this chapter. Traditionally, the study of thermodynamics has emphasized applications to devices such as turbines, pumps, engines, compressors, air conditioners, and so on. The association of thermodynamics with predominantly mechanical devices is extremely restrictive and somewhat unfortunate because it gives a narrow view of

contribution
贡献

preceding
前面的

areas where thermodynamics can be applied. Actually, the principles of thermodynamics apply equally well to other devices of contemporary interest such as solar collectors, MHD (Magneto Hydro Dynamic) generators, rocket engines, fuel cells, wind and wave energy systems, and other systems that transform energy from one form to another. As the basic concepts of thermodynamics unfold, devices used to illustrate the basic principles will be seen to stem from an extremely broad cross section of disciplines. A firm command of thermodynamics, therefore, is essential to practically every phase of an engineer's or scientist's career.

Because thermodynamics deals with such a broad and diverse subject as energy, it is traditionally introduced early in a student's formal education. The principles are expanded in a course in thermodynamics and carry over to courses in fluid mechanics and heat transfer, two other disciplines that, along with thermodynamics, are an integral part of a broader area referred to as the thermal sciences. Thermodynamics also has an important impact on the design of engineering systems, and it plays a major role in the selection of materials as well as in the design methodology of practically all engineering systems.

Before you proceed with the study of thermodynamics, a few words of caution are appropriate. Studying thermodynamics can be compared to constructing a building. The structural integrity of the building can be guaranteed only if the foundation is sound. Similarly, a thorough understanding of thermodynamics can be ensured only if the knowledge of a few underlying principles is sound. The analogy is true of practically all courses in engineering and science, but achieving the results of the analogy in thermodynamics is often complicated by the facts that the introductory material appears to be introduced rather slowly and that the accompanying mathematics is on a very fundamental level. Students often overlook the subtle implications of this introductory material, and they frequently achieve a false sense of security early in the course. They are often tempted to race through the first few chapters before acquiring a firm grasp of the basic concepts. This approach may be successful for a short time, but weaknesses will soon develop as a result of an incomplete understanding of the fundamental and underlying principles.

This chapter begins with definitions of several terms, such as state, process, system and property, which will be used repeatedly throughout the text. The section on definitions is followed by a brief discussion of the system of units that is most commonly encountered in technical fields. The System International (SI) Unit, or international system, is used exclusively in the examples and end-of-chapter problems. The next two sections are devoted to a discussion of pressure and temperature, and the chapter concludes with a discussion of two distinctly different forms of energy transfer: heat transfer and work interactions.

1.2 Carnot Cycle 卡诺循环

The **Carnot cycle** was introduced briefly in the discussion of the Carnot principle and heat engines. In this section the Carnot cycle will be examined in more detail.

卡诺循环是只有两个热源（一个高温热源温度 T_1 和一个低温热源温度 T_2）的简单循环。

A totally reversible cycle such as the Carnot cycle cannot be achieved in practice, because irreversibility accompanies the motion of all fluids and mechanical components. In addition,the heat transfer of the system or from the system can not be reversible due to the limited temperature difference. It is necessary to have a finite amount of heat transfer. However, the study of reversible cycles is instructive because these cycles provide upper limits on the performance of real cycles. The performance of actual heat engines and refrigerators can best be evaluated by comparison with the performance of their reversible counterparts. In addition, improvements in the actual cycle are often deliberate attempts to cause the actual cycles to more nearly approximate the

Carnot cycle
卡诺循环

reversible cycle.

In order for a cycle to be totally reversible, each of the individual processes that make up the cycle must be internally reversible and all heat transfer interactions with the surroundings must occur in a reversible manner. The Carnot cycle operates between two constant-temperature reservoirs and is composed of the following four reversible processes:

1-2: a reversible isothermal expansion during which heat transfer occurs from the high-temperature reservoir to the working fluid

2-3: a reversible adiabatic expansion that continues until the working fluid reaches the temperature of the low-temperature reservoir

3-4: a reversible isothermal compression during which heat transfer occurs from the working fluid to the low-temperature reservoir

4-1: a reversible adiabatic compression that continues until the working fluid reaches the temperature of the high-temperature reservoir

These processes are illustrated with a temperature-entropy **diagram** in Fig.1.1. Notice that since processes 2-3 and 4-1 are both reversible and adiabatic, they are isentropic processes and appear as vertical lines on the *T-s* diagram. The Carnot cycle has a rectangular shape on a *T-s* diagram regardless of the working

diagram
图表

fluid. However, if the cycle is represented on a *p-v* diagram, it can have many different shapes depending on the working fluid and the state of the working fluid in the various parts of the cycle. A typical ideal gas *p-v* diagram for Carnot cycle is shown in Fig. 1.2.

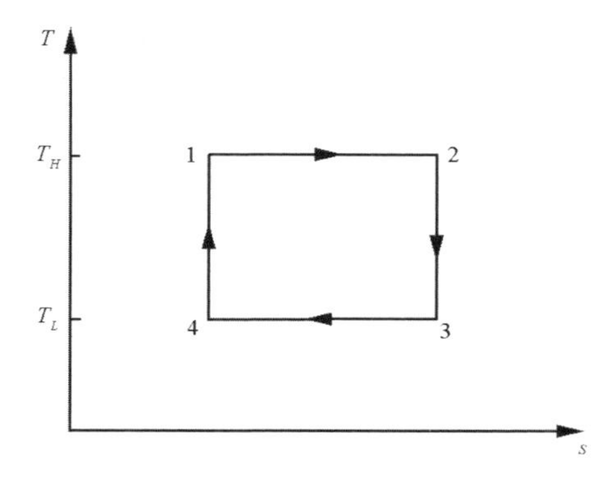

Fig. 1.1 Temperature-entropy diagram for Carnot cycle.

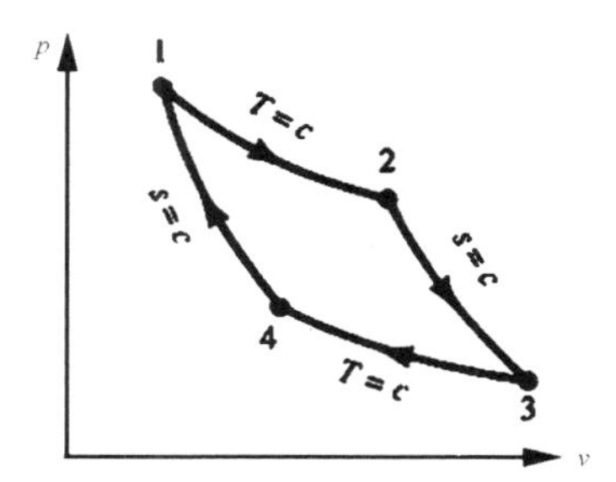

Fig. 1.2 Typical *p-v* diagram of ideal gas for Carnot cycle.

Since the processes that make up the Carnot cycle are reversible, the entropy change associated with each process is given by Equation below:

$$ds = \left(\frac{\delta q}{T}\right)_{\text{intrev}} \tag{1-1}$$

$$\delta q_{\text{intrev}} = T\mathrm{d}s \tag{1-2}$$

and

$$\delta q_{\text{int}\,rev} = Tds \tag{1-3}$$

The physical interpretation of this equation is the heat transfer for each internally reversible process represented by the area under the process curve on the *T-s* diagram. Therefore, for process 1-2 in Fig.1.1,

$$q_{H,\text{intrev}} = T_H(s_2 - s_1) \tag{1-4}$$

and for process 3-4

$$q_{L,\text{intrev}} = T_L(s_4 - s_3) \tag{1-5}$$

and since

$$(s_4 - s_3) = -(s_2 - s_1) \tag{1-6}$$

the ratio of the heat-transfer quantities can be written as

$$\left(\frac{q_L}{q_H}\right)_{\text{intrev}} = \frac{T_L}{T_H} \tag{1-7}$$

The thermal efficiency of the Carnot heat engine is therefore given by

$$\eta_{th,\text{Carnot}} = \frac{q_H - q_L}{q_H} = 1 - \frac{T_L}{T_H} \tag{1-8}$$

These are the same results obtained in previous

discussions of reversible heat engines. The area enclosed by the cycle on the *T-s* diagram represents the net heat transfer for the cycle, since the cycle is internally reversible. The net heat transfer is also equal to the net work of the cycle according to the conservation of energy.

For closed systems the area under the *p-v* diagram represents the *p*d*v* work done by the system during an internally reversible process. Notice, therefore, that a closed system executing a Carnot cycle has work output during the expansion processes 1-2 and 2-3 (Fig.1.2) and work input during the compression processes 3-4 and 4-1.

For open, steady systems *p*d*v* work is not present, since the boundary of the system cannot expand or contract. Therefore, the areas under the process curves on the *p-v* diagram do not represent the work done for the open, steady system. However, in these systems work is done by the system during the reversible adiabatic expansion (process 2-3), and work is done on the system during the reversible adiabatic compression (process 4-1).

If the direction of each process in the Carnot cycle is reversed, it is called reversed Carnot cycle or Carnot refrigerator cycle or Carnot heat-pump cycle. The cycles for the Carnot heat engine and the Carnot refrigerator are used frequently for comparison with actual heat-engine and refrigerator cycles.

1.3 Sensible Heat

显 热

Heat level or heat intensity can readily be measured when it changes the temperature of a substance (remember the example of changing 1 kg of water from 0 to 1 °C). This change in the heat level can be measured with a thermometer.

When a change of temperature can be registered, we know that the level of heat or heat intensity has changed; this is called **sensible heat**.

1.4 Latent Heat

潜 热

潜热，相变潜热的简称，指物质在等温等压情况下，从一个相变化到另一个相吸收或放出的热量。

sensible heat
显热

latent heat
潜热

Another type of heat is called latent or hidden heat. Heat is added in this process, but no temperature rise occurs. An example is heat added to water while it is boiling in an open container. Once water is brought to the boiling point, adding more heat only makes it boil faster; it does not raise the temperature (Fig.1.3).

There are three other terms that are important to understand when referring to **latent heat** transfers: latent heat of vaporization, latent heat of condensation, and latent heat of fusion. Latent heat of vaporization is the amount of heat energy required to change a substance into a vapor. For example, the latent heat of vaporization for water at atmospheric conditions is 2,257.2 kJ/kg. For ease of calculation, we often round this value to 2,257 kJ/kg. This means that if we have 1 kg of water at 100 °C that we want to change to 1 kg of steam at 100 °C, we would need to add 2,257 kJ of heat energy to the water.

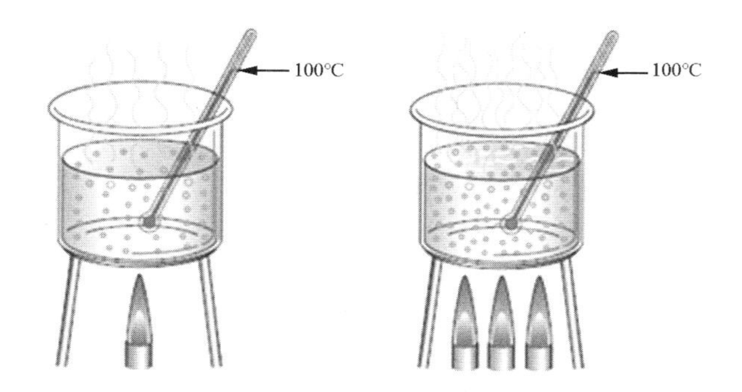

Fig. 1.3 Adding three times as much heat only causes the water to boil faster. The water does not increase in temperature.

Latent heat of condensation is the amount of heat energy, in kJ/kg, required to change a vapor into a liquid. This is, in effect, the opposite of latent heat of vaporization. For example, the latent heat of condensation for steam at atmospheric conditions is 2,257.2 kJ/kg. If we had 1 kg of steam at 100 °C that we wanted to change to 1 lb of water at 100 °C, we would need to remove 2,257 kJ/kg of heat energy from the steam.

Latent heat of fusion is the amount of heat energy needed to change the state of a substance from a solid to a liquid or from a liquid to a solid. The latent heat of fusion for water is 335 kJ/kg. If we had 1 kg of ice at 0 °C under atmospheric conditions, we would have to add 335 kJ/kg of heat energy to the ice in order to melt it. If, on the other hand, we had 1 kg of water at 0 °C under atmospheric conditions, we would have to remove 335 kJ/kg of heat energy from the water in order to freeze it.

1.5 Specific Heat 比热容

比热容，是指没有相变化和化学变化时，1kg均相物质温度升高1K所需的热量。

We now realize that different substances respond differently to heat. When the same amount of heat energy is added to 1 kg of water and gold respectively, the temperatures are increased differently. This difference in heat rise is known as **specific heat**.

specific heat 比热

Specific heat is the amount of heat necessary to raise the temperature of 1 kg of a substance 1 K. Every substance has a different specific heat. Note that the specific heat of water is 4.2 kJ / kg·K. See Tab. 1.1 for the specific heat of some other substances.

Tab. 1.1 The specific heat of several different substances

SUBSTANCE	SPECIFIC HEAT kJ / (kg·K^{-1})	SUBSTANCE	SPECIFIC HEAT kJ / (kg·K^{-1})
Air	1.003	Ice	2.11
Aluminum	0.88	Iron	0.54
Ammonia	2.159	Kerosene	2.09
Beef, lean	3.22	Lead	0.13
Beets	3.77	Marble	0.817
Building Brick	0.837	Porcelain	0.754
Concrete	0.653	Pork, fresh	2.85
Copper	0.385	Seawater	3.89
Cucumbers	4.06	Shrimp	3.48
Eggs	3.18	Spinach	3.94
Fish	3.18	Steam	2.09
Flour	1.59	Steel	0.46
Gold	0.129	Water	4.19

本章重点内容介绍

热力学是一门物理科学，它的大部分研究主题是能量及其转换。研究热力学的基本原理是质量守恒、能量守恒和热力学第二定律。热力学的一个主要研究方面是转移到一种物质上的能量和该物质性质变化之间的数学关系。

如果卡诺循环中每个过程的方向是相反的，那么产生的循环称为反向卡诺循环、卡诺制冷循环或卡诺热泵循环。卡诺热机和卡诺制冷机的循环经常用于与实际热机和制冷机循环进行比较。

当物质的温度发生变化时，可以很容易地测量热水平或热强度的变化（记住将 1 kg 的水从 0℃变为 1℃的例子）。这种热量的变化可以用温度计来测量。

当温度的变化被记录下来时，我们知道热的水平或热的强度已经改变了，这叫作显热。在这个过程中加入了热量，但温度没有升高。比如当水在一个开放的容器中沸腾时，将其加热，一旦水达到沸点，加入更多的热量只会使它沸腾得更快，但不会提高温度。

当提到潜热传递时，还有三个术语需要理解：汽化潜热、冷凝潜热和熔化潜热。汽化潜热是将一种物质由液态转变为蒸气所需的热能，单位为 kJ/kg。冷凝潜热是将蒸气转变为液体所需的热能，单位为 kJ/kg。熔化潜热是将物质从固态变为液态或从液态变为固态所需的热能，单位为 kJ/kg。

当相同的热能分别加入 1 kg 的水和金中时，温度会有不同的升高。这种热升的差异被称为比热。

Chapter 2
Heat Transfer
传热学

2.1 Introduction
简　介

Heat transfer is a **branch** of applied thermodynamics. It may be defined as the analysis of the rate at which heat is transferred across system boundaries subjected to specific temperature difference conditions. Whereas classical thermodynamics deals with the amount of heat transferred during a process, heat transfer estimates the rate at which heat transfers and the temperature distribution of the system during the process. We are interested in the **transient**, non-equilibrium exchange process; not just the **static** equilibrium states before and after the process.

传热学，是研究由温差引起的热能传递规律的科学。

branch
分支

transient
瞬态的

static
静态的

Consider a hot sphere quenched by immersion in cold water. Thermodynamics tells us what the equilibrium states of sphere and water will be after the process. Heat transfer enables us to analyze the temperature distributions in both sphere and water and the transient heat exchange during the process. It may be a complex analysis — for example the water may boil or be in streaming motion, the sphere may be anisotropic or layered — but nevertheless the time-varying temperatures and exchange rates are the goals of a heat transfer analysis.

The foundation blocks of heat transfer analysis are the first and second laws of thermodynamics. The second law tells us that heat flows from high to low temperature, that is, in the direction of decreasing temperature. Certain limitations on maximum and minimum temperatures and on overall system efficiency are also required by the second law. Otherwise the second law does not directly intrude upon a heat transfer analysis.

In contrast, the first law of thermodynamics is the fundamental relation behind every heat transfer analysis. It may be stated either for a closed system or for an open control volume. In system form, the total energy increase of the system equals the heat received plus the work received:

$$\frac{\mathrm{d}E}{\mathrm{d}t}=\frac{\mathrm{d}Q}{\mathrm{d}t}+\frac{\mathrm{d}W}{\mathrm{d}t} \qquad (2\text{-}1)$$

Some authors define work received by the system as negative. For conduction in solids, the work term,, is negligible. For convection in fluids,

dW/dt the work done by fluid pressure forces is important, but the changes in fluid kinetic and potential energy are usually small compared to the heat transfer rate, dQ/dt.

For steady flow at low speed through a fixed control volume, the first law of thermodynamics takes the form of a control surface integral:

$$\frac{dQ}{dt} = \int_{CS} \int i \rho V_n dA \qquad (2\text{-}2)$$

where $i = e + p/\rho$ is the fluid **enthalpy** and v_n is the fluid velocity normal to the surface, taken as positive if exiting and negative if entering. We assume no shaft work in Eq. (2-2). In very high speed gas flows, the fluid kinetic energy and dissipative viscous work should be retained.

Equation (2-1) or (2-2) is the basis for practically every analysis in this text. The many different kinds of shaft work and shear work occurring in thermodynamics are rarely important here, so heat transfer is somewhat simpler in principle than thermodynamics, a fact that you may appreciate.

Both equations have a defect, however: They do not provide explicit information about the heat transfer rate, dQ/dt. It is the task of heat transfer analysis to provide additional equations, either theoretical or empirical, t_0 relate dQ/dt to system parameters such as geometry, material properties,

enthalpy
焓

flow rates, and environmental temperatures. The purpose of this text is thus to supplement the first law of thermodynamics with accurate and physically plausible heat transfer relationships. Additional data about the thermodynamic state are also required: density and specific heats and, if **phase changes** occur, the latent enthalpies of melting, vaporization, and sublimation. Problems involving fluid flow may require still other properties such as viscosity, surface tension, and coefficient of thermal expansion.

Heat transfer is concerned with temperature differences, and we live in a world full of such differences, owing to either natural or artificial causes. Thus an immense variety of heat transfer equipment has been created to deal with these differences: boilers, condensers, solar collectors, radiators, compact heat exchangers, combustion chambers, furnaces, driers, distillation columns, refrigerators, insulators, stoves — the list is almost endless. Each application requires the first law of thermodynamics plus material data and heat transfer relations appropriate to the specific system.

A familiar example is the truck radiator shown in Fig. 2.1. This is a type of compact heat exchanger. The fluid entering the radiator has been heated in another heat exchanger surrounding the combustion chambers of the truck engine. It must be cooled in the radiator by

phase changes
相变

air passing over multi-finned surfaces. The fins enhance the exchange by providing hotter surface for the air to cool. The cooler exit fluid then returns to the truck engine to absorb more heat from the combustion surfaces. It is the task of the heat transfer analyst to ensure that this radiator is economical and reliably efficient over all practical operating ranges of engine power, coolant liquid flow, and air flow. This text develops accurate and plausible methods of analyzing such problems.

Fig. 2.1 A finned, drawn-tank radiator.

2.2 Modes of Heat Transfer
传热的方式

Given the fact from the second law of thermodynamics that heat flows across temperature differences, how many ways can heat be transferred? As you probably know from experience, there are three modes of heat transfer: conduction, radiation, and convection. Conduction and radiation are fundamental

传热的方式：导热，辐射，对流。

physical **mechanisms**, while convection is really conduction as affected by fluid flow.

Conduction is an exchange of energy by direct interaction between molecules of a substance containing temperature differences. It occurs in gases, liquids, or solids and has a strong basis in the molecular kinetic theory of physics.

Radiation is a transfer of thermal energy in the form of electromagnetic waves emitted by atomic and **subatomic** agitation at the surface of a body. Like all electromagnetic waves (light, X-rays, microwaves), thermal radiation travels at the speed of light, passing most easily through a vacuum or a nearly "transparent" gas such as oxygen or nitrogen. Liquids, "participating" gases such as carbon dioxide and water vapor, and glasses transmit only a portion of incident radiation. Most other solids are essentially opaque to radiation. The analysis of thermal radiation has a strong theoretical basis in physics, beginning with the work of Maxwell and of Planck.

Convection may be described as conduction in a fluid as enhanced by the motion of the fluid. It may not be a truly independent mode, but convection is the most heavily studied problem in heat transfer: More than three-quarters of all published heat transfer papers deal with convection. This is because convection is a difficult subject, being strongly influenced by

mechanism
机制

subatomic
亚原子的

geometry, turbulence, and fluid properties.

Let us look at each mode in somewhat more detail. A given problem may, of course, involve two or even all three modes.

2.3 Heat Conduction 热传导

Conduction transfers energy from hot to cold regions of a substance by molecular interaction. In fluids, the exchange of energy is primarily by direct impact. In solids, the primary mechanism is relative lattice vibrations, enhanced in the case of metals by drift of free electrons through the lattice. Thus good electrical conductors are also good heat conductors. Both the molecular and the free-electron interactions are well founded in theoretical atomic physics.

导热是介质内无宏观运动时的传热现象，其在固体、液体和气体中均可发生。

As engineers, we wish to know on a **macroscopic** level how much heat is transferred by conduction. How can we express such a low using macroscopic variables such as temperature? The following homely example may help.

Consider the insulation in the wall of a home. Neglect the plaster and wood frame and let the thick insulation be the dominant effect on conduction through the wall. What affects heat loss from the warm room to the cold outside?

macroscopic
宏观的

First, we know that the cooler it is outside, the greater the heat loss. The bigger the house is, the greater the heat loss. But the thicker the insulation, the less the heat loss: Advertisements tell us that doubling the insulation halves the loss. Thus we deduce three effects characteristic of conduction heat flow and can write a **qualitative** expression:

$$\text{Conduction heat flow} \propto (\text{Wall area})\frac{(\text{Temperatue difference})}{(\text{Wall thickness})} \quad (2\text{-}3)$$

The proportionality "constant" is a property of the material (in this case the insulation).

These deductions were made more than two centuries ago by Joseph Fourier, a French mathematical physicist, and published in 1822 in his pioneering heat transfer book. Fourier wrote Eq. (2-3) in limiting mathematical form, assuming that temperature T varies in the x direction, say, and using the short notation $q_x = \mathrm{d}Q/\mathrm{d}t$ in the x direction:

$$q_x = -kA_x \frac{\mathrm{d}T}{\mathrm{d}x} \quad (2\text{-}4)$$

or

$$q'' = \frac{q_x}{A_x} = -k\frac{\mathrm{d}T}{\mathrm{d}x} \quad (2\text{-}5)$$

qualitative
定性的

Fourier's Law of Conduction
傅立叶导热定律

This is **Fourier's law of heat conduction**; it is valid for all common solids, liquids, and gases. The coefficient k is a material transport property called the thermal conductivity. The double-prime notation, $q''=q/A$, will be convenient in many analyses. The quantity A_x is the area normal to the x direction through which the heat flows.

2.4 Radiation
辐 射

Unlike conduction, which requires a medium, radiation is an electromagnetic phenomenon and travels easily through a vacuum at the speed of light. Most gases transmit nearly all incident radiation, but liquids, even clean water, rapidly attenuate radiation. Most solids, except for glasses and clear plastics, are completely opaque to radiation.

辐射是物体用电磁辐射的形式把热能向外散发的传热方式。

All opaque surfaces emit thermal radiation and absorb or reflect incident radiation. A perfect or "blackbody" surface emits at a maximum rate and, correspondingly, absorbs all incident radiation. The term blackbody derives from the fact that a black surface absorbs all incident radiation in the visible range and thus reflects no colors to the eye. The net rate of heat flux from an opaque surface equals the total energy emitted and reflected minus the total energy absorbed from the surroundings.

Experiments by Stefan in 1879 and a theory by L. Boltzmann in 1884 showed that a black surface emits radiant energy at a rate-proportional to the fourth power of the absolute temperature of the surface. If a black surface has area A and temperature T, its radiant emission is given by

$$E_b = \sigma A T^4 \tag{2-6}$$

Where σ is a fundamental proportionality called the Stefan-Boltzmann constant. It is related to Planck's constant and Boltzmann's constant and has a numerical value of

$$\sigma = 5.67 \times 10^{-8}\ \mathrm{W/(m^2 \cdot k^4)} \tag{2-7}$$

Note that E_b in Eq. (2-4) has units of heat flux: W.

Real surfaces are nonblack and emit radiation at a rate less than maximum. A convenient way to express this is to say that they emit at a fraction, E, of the blackbody rate. For a real surface:

$$E = \varepsilon E_b = \varepsilon \sigma A T^4 \tag{2-8}$$

The dimensionless parameter ε is called the emissivity of the surface and varies between zero and unity. Experiments show that ε varies with temperature and also with surface parameters: roughness, texture, color, degree of oxidation, and the presence of coatings. An idealized material with constant emissivity is called a gray body.

2.5 Convection
对 流

The dictionary defines convection as "the conveying of heat through a liquid or gas-by motion of its parts." This is good, but how does it happen, what is the mechanism?

对流是指由于流体的宏观运动而引起的流体各部分之间发生相对位移，冷热流体相互掺混所引起的热量传递过程。

First, we require a fluid, which may be a liquid or gas or a mixture (for example, boiling or condensation). And the fluid must be in motion or else things revert to a static conduction type of situation, such as in Fig.2.2. For heat flux in Fig.2.2, we could have added some static fluid examples such as water (q_x''=115 W/m^2) or air (q_x'' =5 W/m^2). But we would have to justify that no motion enhances, because motion enhances the heat flux.

Second, we usually have a solid surface next to the fluid. Granted there are cases of convection involving only fluids, such as a hot jet **discharging** into a cold reservoir, but the majority of practical problems involve a hot (or cold) surface adding (or subtracting) heat to (or from) the fluid.

discharging
排放，释放

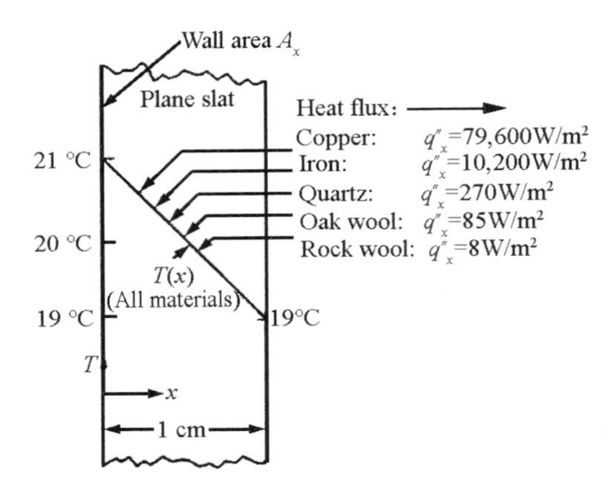

Fig. 2.2 Temperature distribution and conduction heat flux in a plane slab for various solid materials.

Finally, we distinguish between forced versus free convection, as illustrated in Fig.2.3. In forced convection (Fig.2.3[a]), there is a dominant streaming motion far from the solid surface and of independent cause, such as a fan or a pump or the wind. In free convection (Fig.2.3[b]), far field streaming is negligible and the only significant motion is caused by differential **buoyancy** adjacent to the hot (cold) surface. The heat flux transferred from the solid is usually much stronger for forced convection and increases with magnitude of the streaming velocity. Finally, if the far field streaming is relatively minor, such as the small fan in Fig. 2.3(c), the effects of streaming and buoyancy will be comparable. This is called mixed convection, and the convection heat flux will be intermediate or moderate in magnitude.

buoyancy
浮力

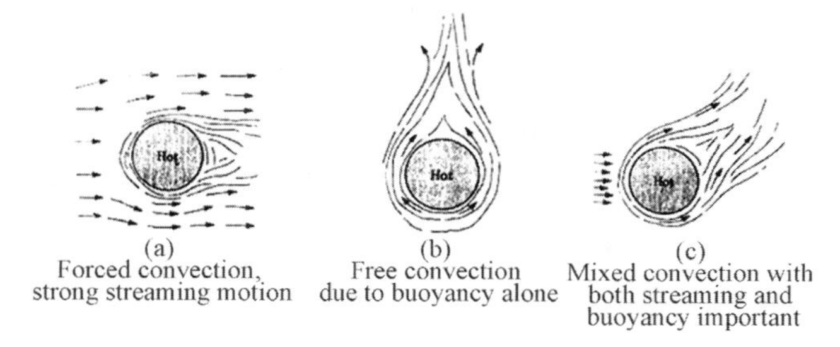

Fig. 2.3 Convection regimes for flow near a hot sphere.

本章重点内容介绍

传热是应用热力学的一个分支。它可以定义为在特定的温差条件下，对热量跨系统边界传输速率的分析。经典热力学处理的是过程中的传热量，而传热则估计的是过程中的传热速率和系统的温度分布。我们不仅仅研究过程前后的静态平衡状态，还研究瞬态、非平衡交换过程。

传热分析的基础是热力学第一定律和第二定律。第二定律告诉我们，热量从高温流向低温，即朝着降温的方向流动。第二定律还要求对最高和最低温度以及整个系统效率进行某些限制。否则，第二定律不会直接影响传热分析。热力学第一定律是每一个传热分析背后的基本关系。

传热分析的任务是提供与系统参数（如几何结构、材料特性、流速和环境温度）相关的附加方程（理论或经验）。因此，本章的目的是补充热力学第一定律与物理上的传热关系。还需要关于热力学状态的额外数据：密度和比热，以及如果发生相变，熔化、汽化和升华的潜热。涉及流体流动的问题可能还需要其他特性，如黏度、表面张力和热膨胀系数。

热传递与温度差异有关，我们生活在一个充满这种差异的世界里，这是由于自然或人为原因造成的。因此，为了解决这些差异，人们发明了各种各样的传热设备，每种应用都需要热力学第一定律加上适合于特定系统的材料数据和传热关系。

传导是通过含有温差的物质分子间的直接相互作用进行的能量交换。辐射是一种热能的转移，其形式是由物体表面原子和亚原子的搅动所发出的电磁波。对流可以被描述为流体中的传导，其被流体的运动增强。对流是传热学中研究最多的问题。一个给定的问题可能涉及两种甚至全部三种模式。

Chapter 3
Fluid Mechanics
流体力学

3.1 Introduction
简　介

A proper understanding of fluid mechanics is extremely important in many areas of engineering. In biomechanics the flow of blood and cerebral fluid are of particular interest; in meteorology and ocean engineering an understanding of the motions of air movements and ocean currents requires a knowledge of the mechanics of fluids, chemical engineers must understand fluid mechanics to design the many different kinds of chemical-processing equipment; aeronautical engineers use their knowledge of fluids to maximize lift and minimize drag on aircraft and to design fan-jet engines; mechanical engineers design pumps, turbines, internal combustion engines, air compressors, air-conditioning, pollution-control equipment, and power plants using a proper understanding of fluid mechanics;

流体力学主要研究在各种力的作用下，流体本身的静止状态和运动状态以及流体和固体界壁间有相对运动时的相互作用和流动规律。

and civil engineers must also utilize the results obtained from a study of the mechanics of fluids to understand the transport of river sediment and erosion, the pollution of the air and water, and to design piping systems, sewage treatment plants, irrigation channels, flood control systems, dams, and domed athletic stadiums. Fig.3.1 displays the experiment in the tunnel where the flow field could be visualized clearly to understand the fluid mechanism.

Fig. 3.1 Experiment in wind tunnel.

Since the earth is almost 75% covered with water and 100% covered with air, the scope of fluid mechanics is vast and touches nearly every human endeavor. It is not possible to present fluid mechanics in such a way that all of the foregoing subjects can be treated specifically; it is possible, however, to present the fundamentals of the mechanics of fluids so that engineers are able to understand the role that the fluid plays in a particular application. This role may involve the proper sizing of a pump (the horsepower and flow rate) or the calculation of a force acting on a structure.

3.2 Basic Fluid Mechanics
流体力学基础

边界层是高雷诺数绕流中紧贴物面的黏性力不可忽略的流动薄层。

In dealing with convection problems it is important to have an understanding of the behavior of fluids in motion over external surfaces or through enclosed channels. For liquids or gases flowing over external surfaces under continuum conditions, it will be found that the relative velocity between the surface and the fluid to zero at the surface. Moving away from the surface, the velocity increases rapidly toward the free-stream value, effectively reaching the free-stream value at a distance S not far from the surface. The thin region where the velocity is varying is called the boundary layer (Fig.3.2),a term suggested by Prandtl who first recognized this basic phenomenon. Since the shearing stress is proportional to the product of the viscosity and the velocity gradient, it is clear that substantial shearing stresses will occur only in the boundary layer where a velocity gradient exists, whereas outside the boundary the shearing stress will be vanishingly small. Accordingly, it may be stated that the effects of viscosity are confined to the boundary layer, whereas outside the boundary layer the flow may be considered to be **inviscid**. Thus in analyzing the flow field over external surfaces, the inviscid flow equations may be used to predict the free-stream flow field. The resulting velocity distribution may then be used

inviscid
无粘性的

in conjunction with the boundary layer equations which include the influence of viscosity for the prediction of the flow field in the immediate vicinity of the wall. In this manner the drag on external surface can be determined.

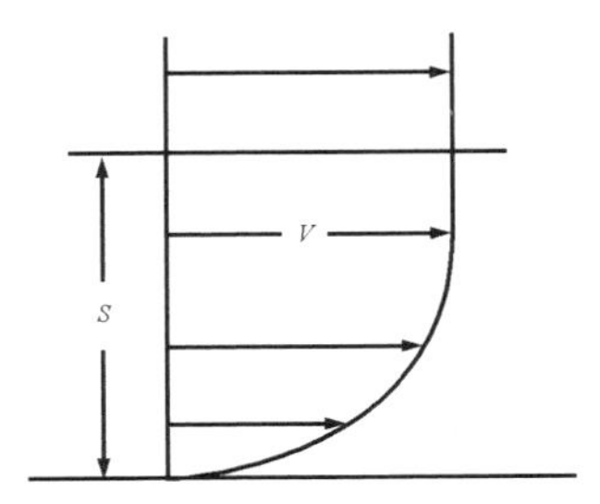

Fig. 3.2 The boundary layer.

In the case of heat transfer (or mass transfer) to or from external surfaces placed in a flow field, it will be found that there is a thermal boundary layer (or concentration boundary layer) analogous to the velocity boundary layer, within which the influence of thermal conductivity (or diffusivity) is confined. Outside this region the flow is essentially non-conducting and non-diffusing.

In the case of a fluid flow through an enclosed channel (Fig.3.3), a boundary layer begins at the channel entrance. In this entrance region there is an inviscid core flow and a viscous boundary layer. Some distance downstream the boundary layers grow together and the velocity is at a maximum in the central region of the duct, decreasing to a value of zero at the bounding surfaces.

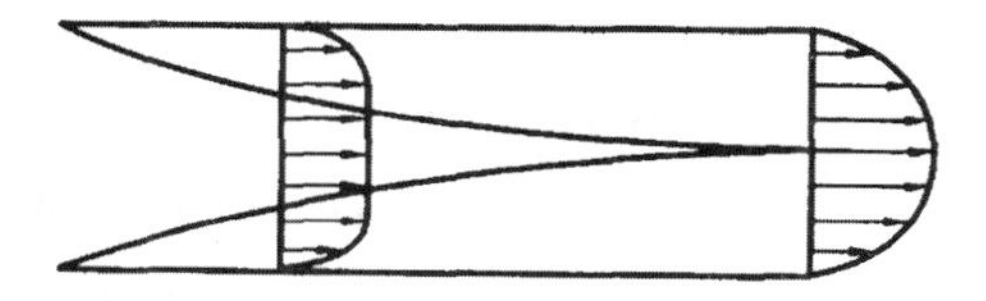

Fig. 3.3 A flow through an enclose channel.

(1) Laminar and Turbulent Flows

Osborne Reynolds in 1883 reported that there are two basically different types of fluid motion which he identified as **laminar** flow and **turbulent** flow. For example, in the case of flow over a flat plate geometry the boundary layer motion near the leading edge of the plate is smooth or streamlined. Locally within the boundary layer the velocity is constant and invariant with time.

In this region momentum and energy transfer occur by a **diffusion** process as described by the Newtonian shearing stress law and by the Fourier conduction relationship. This is the region of laminar flow. If the plate is long enough or the velocity sufficiently and we proceed far downstream, the nature of the flow is markedly changed (Fig.3.4). At any point in the boundary layer the velocity varies with time about some mean value as shown in Fig.3.5. The exchange of momentum and energy is now no longer controlled by diffusional processes. Rather macroscopic eddies randomly move from one fluid layer to another, and in the process

laminar
层流

turbulent
湍流

diffusion
扩散

momentum and energy are transferred. The analysis of transport processes in turbulent flows is inherently more difficult than in the laminar cases and, in general, the treatment is semi-empirical in nature.

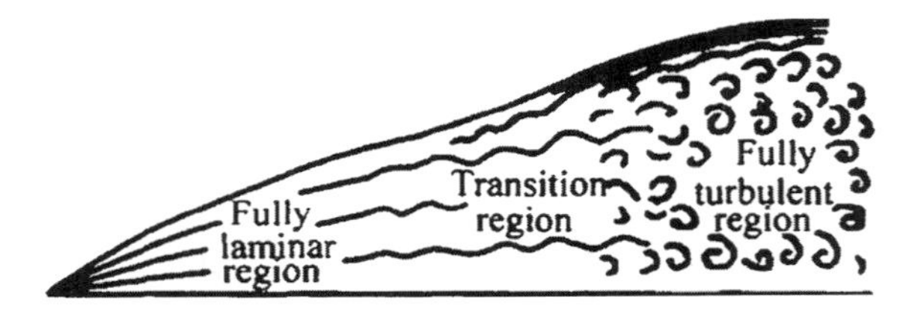

Fig. 3.4 The change of laminar flow into turbulent flow.

The flow does not change abruptly laminar to turbulent motion, but rather there is an intermediate region connecting the well defined laminar and the well defined turbulent motion. This is the transition region. It has been found that the laminar boundary layer begins to experience transition where the **dimensionless** quality ($u_e x / v$) , called the critical Reynolds number for flow over external surfaces, is of the order of 500,000, but this is dependent on the level of turbulence in the free stream.

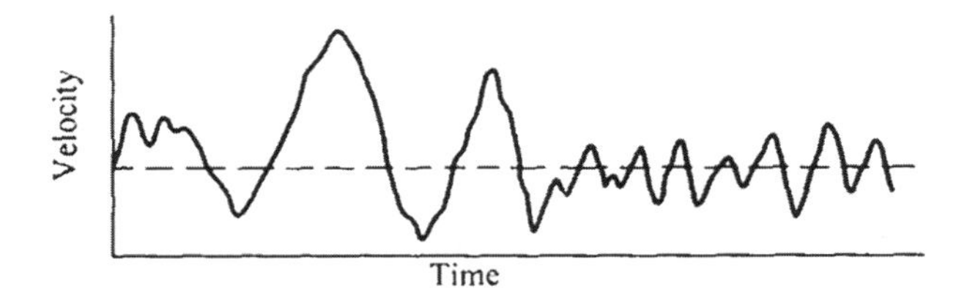

Fig. 3.5 The relationship between velocity and time in turbulent flow.

dimensionless
无量纲的

For flow in circular tubes, it has been found that the flow is generally laminar if the Reynolds number $\bar{u}d/v$, where $\bar{u}$ is mean velocity, d is pipe diameter, and v is kinematic viscosity, is lower than 2,300. If this Reynolds number is greater than 10,000, the flow is considered to be fully turbulent. In the 2,300 to 10,000 region, the flow is described as transition flow. It is possible to shift these Reynolds values by minimizing the disturbances in the inlet flow, but for general engineering applications the numbers cited are representative.

(2) Flow Separation

In region of adverse pressure gradient such as encountered in flow over curved bodies, the boundary layer, in effect, separates from the surface. At this location the shear stress goes to zero and beyond this point there is a reversal of flow in the vicinity of the wall. In this separated the region, the boundary layer equations are no longer valid and the analysis of the flow is generally very difficult.

3.3 Conservation Law 守恒定律

From experience it has been found that fundamental laws exist that appear exact; that is, if experiments are conducted with the utmost precision and care, deviations from these laws are very small and in fact, the deviations would be even smaller if improved experimental techniques were employed. Three such laws form the basis for our study of fluid mechanics. The first is the conservation of mass, which states that matter is indestructible. Even though Einstein's theory of relativity postulates that under certain conditions, matter is convertible into energy and leads to the statement that the extraordinary quantities of radiation from the sun are associated with a conversion of 3.3×10^{14} kg of matter per day into energy, the destructibility of matter under typical engineering conditions is not measurable and does not violate the conservation of mass principle.

For the second and third laws it is necessary to introduce the concept of a system. A system is defined as a fixed quantity of matter upon which attention is focused. Everything external to the system is separated by the system boundaries. These boundaries may be fixed or movable, real, or imagined. With this definition we can now present our second fundamental law, the conservation of momentum: The momentum of a system remains constant if no external forces are acting on the system. A more specific law based on this principle is Newton's second law: The sum of all external forces acting on a system is equal to the time rate of change of linear momentum of the system. A parallel law exists for the moment of momentum: The rate of change of angular momentum is equal to the sum of all torques acting on the system.

The third fundamental law is the conservation of energy, which is also known as the first law of thermodynamics: The total energy of an isolated system remains constant. If a system is in contact with the surroundings, its energy increases only if the energy of the surroundings experiences a corresponding decrease. It is noted that the total energy consists of potential, kinetic, and internal energy, the latter being the energy content due to the temperature of the system. Other forms of energy are not considered in fluid mechanics.

3.4 Computational Fluid Dynamics 计算流体动力学

计算流体动力学是近代流体力学、数值数学和计算机科学结合的产物，是一门具有强大生命力的交叉科学。

The equations governing fluid-flow problems are the continuity, the Navier-Stokes, and the energy equations. These equations form a system of coupled quasi-linear partial differential equations (PDEs). Because of the nonlinear terms in these PDEs, analytical methods can yield very few solutions. In general, analytical solutions are possible only if these PDEs can be made linear, either because nonlinear terms naturally drop out (e.g., fully developed flows in ducts and flows that are irrotational everywhere) or because nonlinear terms are small when compared to other terms so that they can be neglected(e.g., flows where the Reynolds number is less than unity). Yih and Schlichting & Gersten describe most of the better-known analytical solutions. If the nonlinearities in the governing PDEs cannot be neglected, which is the situation for most engineering flows, then numerical methods are needed to obtain solutions.

Computational fluid dynamics, or simply CFD, is concerned with obtaining numerical solutions to fluid-flow problems by using the computer. The advent of high-speed and large-memory computers has enabled CFD to obtain solutions to many flow problems, including those that are compressible or incompressible, laminar or turbulent, chemically reacting or nonreacting, single- or multi-phase. Of the numerical methods developed to address equations governing fluid-flow problems, finite-difference methods (FDMs) and finite-volume methods (FVMs) are the most widely used.

Since computers are used to obtain solutions, it is important to understand the constraints that they impose. Of these constraints, four are critical. The first of these is that computers can perform only arithmetic (i.e., add, subtract, multiply and divide) and logic (i.e., true and false) operations. This means non-arithmetic operations, such as derivatives and integrals, must be represented in terms of arithmetic and logic operations. The second constraint is that computers represent numbers using a finite number of digits. This means that there are round-off errors and that these errors must be controlled. The third constraint is that computers have limited storage memories. This means solutions can be obtained only at a finite number of points in space and time. Finally, computers perform a finite number of operations per unit time. This means solution procedures should minimize the computer time needed to achieve a computational task by fully utilizing all available processors on a computer and minimizing the number of operations.

With these constraints, FD methods generate solutions to PDEs through the following three major steps:

(1) Discretize the domain. The continuous spatial and temporal domain of the problem must be replaced by a discrete one made up of grid points or cells and time levels. The ideal discretization uses the fewest number of grid points/cells and time levels to obtain solutions of desired accuracy.

(2) Discretize the PDEs. The PDEs governing the problem must be replaced by a set of algebraic equations with the grid points/cells and the time levels as their domain. Ideally, the algebraic equations—referred to as finite-difference equations (FDEs) should describe the same physics as those described by the governing PDEs.

(3) Specify the algorithm. The step-by-step procedure by which solutions at each grid points/cells are obtained from the FD equations when advancing from one time level to the next must be described in detail. Ideally, the algorithm should ensure not only accurate solutions but also efficiency in utilizing the computer.

本章重点内容介绍

正确理解流体力学在许多工程领域是非常重要的。土木工程师还必须利用流体力学研究的结果来了解河流沉积物和侵蚀的运输、空气和水的污染，并设计管道系统、污水处理厂、灌溉渠道、防洪系统、水坝和圆顶体育场。在处理对流问题时，必须了解流体在外部表面或通过封闭通道运动的行为。对于在连续体条件下流过外部表面的液体或气体，表面与流体之间的相对速度在表面为零。由于剪切应力与黏度和速度梯度的乘积成正比，很明显，只有在存在速度梯度的边界层才会发生大量的剪切应力，而在边界外，剪切应力将变小。因此，可以说，黏度的影响仅限于边界层，而在边界层之外，流动可能被认为是无黏的。

流场中物体外表面的传热（或传质）存在类似于速度边界层的热边界层(或浓度边界层)，其中限制了热导率(或扩散率)的影响。在这个区域之外，流动本质上是不导电和不扩散的。

在流体流过封闭通道的情况下，边界层从通道入口开始。在这个入口区域有一个无黏性的岩心流动和一个黏性的边界层。速度在边界层下游的某一处逐渐增大，在管道中心区域达到最大，在边界表面降至零。

奥斯本·雷诺兹在1883年报道了两种基本不同类型的流体运动——层流和湍流。例如，在平板几何形状上的流动，板前缘附近的边界层运动是光滑的或流线型的。在局部边界层内，速度是恒定的，不随时间改变。

在这个区域中，动量和能量传递是由牛顿剪应力定律和傅里叶传导关系描述的扩散过程发生的。动量和能量的交换现在不再受扩散过程的控制。相反，宏观涡流从一个流体层随机移动到另一个流体层，在这个过程中，动量和能量被转移。

对于圆管中的流动，如果雷诺数低于2 300，则流动一般是层流。如果这个雷诺数大于10 000，则认为流动是完全湍流。在2 300到10 000的区域，流动被描述为过渡流。在不利压力梯度的区域，如在弯曲物体上遇到的流动，边界层实际上与表面分离。在这个位置，剪切应力为零，超过这一点，在壁面附近有一个流动反转。在这个分离的区域中，边界层方程不再有效，对流动的分析通常是非常困难的。

由于地球几乎75%被水覆盖，100%被空气覆盖，流体力学的范围是巨大的，几乎涉及人类的每一项努力。在生物力学中，血液和脑液流特别令人感兴趣；在气象学和海洋工程中，了解空气运动和洋流的运动则需要了解流体力学，化学工程师必须了解流体力学，以设计许多不同类型的化学处理设备；航空工程师利用其流体知识，最大限度地提高飞机升力和最大限度地减少飞机阻力，并设计风扇喷气发动机；

机械工程师使用对流体力学的适当理解来设计泵、涡轮机、内燃机、空气压缩机、空调、污染控制设备和发电厂。

Chapter 4
Refrigeration
制　冷

4.1 Introduction
简　介

Refrigeration was used by ancient civilizations when it was naturally available. The Roman rulers had slaves transport ice and snow from the high mountains to be used to preserve foods and to provide cool beverages in hot weather. Such natural sources of refrigeration were, of course, extremely limited in terms of location, temperatures, and scope. Means of producing refrigeration with machinery, called mechanical refrigeration, began to be developed in the 1850s. Today the refrigeration industry is a vast and essential part of any technological society, with yearly sales of equipment amounting to billions of dollars in the United States only.

制冷是应用热力学原理，用人工制造并保持低温的方法。

The following example uses a **residential** window air conditioning system to explain

residential 住宅的

the basics of refrigeration. Residential air-conditioning, whether a window unit or a central system, is considered to be high-temperature refrigeration and is used for comfort cooling.

The refrigeration concepts utilized in the residential air conditioner are the same as those in the household refrigerator. The air conditioner pumps the heat from inside the house to the outside of the house, just as the household refrigerator pumps heat from the refrigerator into the kitchen. In addition, just as heat leaks into the refrigerated compartments in the refrigerator, heat leaks into the house and must be removed. The heat is transferred outside by the air-conditioning system. The cold air in the house is recirculated air. Room air at approximately 23.9 °C goes into the air-conditioning unit, and air at approximately 12.8 °C comes out. This is the same air with some of the heat removed (Fig. 4.1).

Fig.4.1 and the following statements illustrate this concept and are also guidelines to some of the design data used throughout the air-conditioning field.

(1) The outside design temperature is 35 °C.

(2) The inside design temperature is 23.9 °C.

(3) The design cooling coil temperature is 4.4 °C. This coil transfers heat from the room

into the refrigeration system. Notice that with a 23.9 °C room temperature and a 4.4 °C cooling coil temperature, heat will transfer from the room air into the refrigerant in the coil.

(4) The heat transfer makes the air leaving the coil and entering the fan about 12.8 °C. The air exits the fan also at 12.8 °C.

(5) The outside coil temperature is 51.7 °C. The coil transfers heat from the system to the outside air. Notice that when the outside air temperature is 35 °C and the coil temperature is 51.7 °C, heat will be transferred from the system to the outside air.

Careful examination of Fig. 4.1 shows that heat from the house is transferred into the refrigeration system through the inside coil and transferred to the outside air from the refrigeration system through the outside coil. The air-conditioning system is actually pumping the heat out of the house. The system capacity must be large enough to pump the heat out of the house faster than it is **leaking** back in so that the occupants will not become uncomfortable.

leaking
泄露

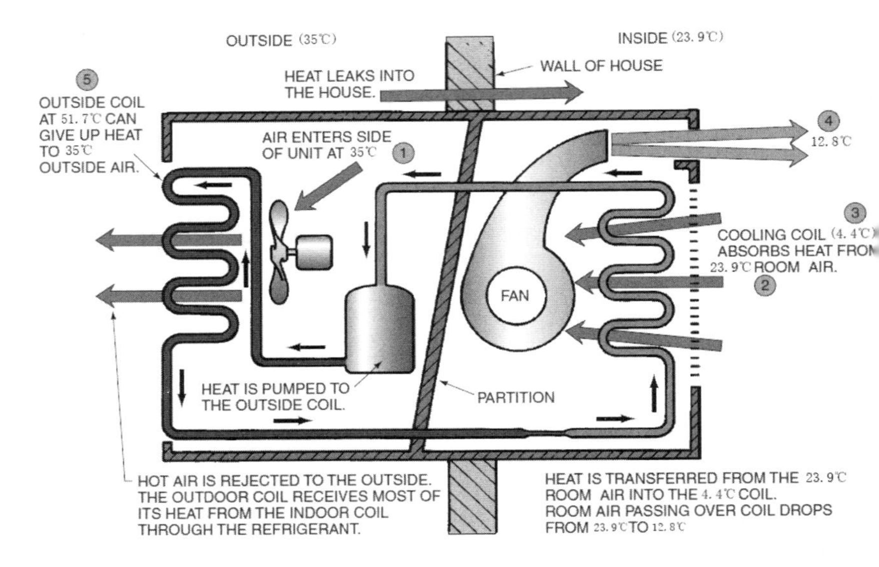

Fig. 4.1 A window air-conditioning unit.

4.1.1 Uses of Refrigeration
制冷的用途

It is convenient to classify the applications of refrigeration into the following categories: domestic, commercial, industrial, and air conditioning. Sometimes transportation is listed as a separated **category**. Domestic refrigeration is used for food preparation and preservation, ice making, and cooling beverages in the household. Commercial refrigeration is used in retail stores, restaurants, and institutions, for purposes the same as those in the household. Industrial refrigeration in the food industry is needed in processing, preparation, and large-scale preservation. This

category
类别

includes use in food chilling and freezing plants, cold storage warehouses, breweries, and dairies, to name a few. Hundreds of other industries use refrigeration; among them are ice-making plants, oil **refineries**, **pharmaceuticals**. Of course ice skating rinks need refrigeration.

Refrigeration is also widely used both in comfort air conditioning for people and in industrial air conditioning. Industrial air conditioning is used to create the air temperatures, humidity, and cleanliness required for manufacturing processes. Computers require a controlled environment.

4.1.2 Method of Refrigeration 制冷的方法

Refrigeration, commonly spoken of as a cooling process is more correctly defined as the removal of heat from a substance to bring it to or keep it at a desirable low temperature, below the temperature of the surroundings. The most widespread method of producing mechanical refrigeration is called the vapor compression system. In this system a volatile liquid refrigerant is evaporated in an evaporator; this process results in a removal of heat (cooling) from the substance to be cooled. A compressor and condenser are required to maintain the evaporation process and to recover the refrigerant for reuse.

refineries
炼油厂

pharmaceuticals
药品

Another widely used method is called the absorption refrigeration system. In this process a refrigerant is evaporated (as with the vapor compression system), but the evaporation is maintained by absorbing the refrigerant in another fluid.

Other refrigeration methods are thermoelectric, steam jet, and air cycle refrigeration. These systems are used only in special applications and their functioning will not be explained here. Thermoelectric refrigeration is still quite expensive: some small tabletop domestic refrigerators are cooled by this method. Steam jet refrigeration is inefficient. Often used on ships in the past, it has been largely replaced by the vapor compression system. The air cycle is sometimes used in air conditioning of aircraft cabins. Refrigeration at extremely low temperatures, below about –130 °C, is called cryogenics. Special systems are used to achieve these conditions. One use of refrigeration at ultra-low temperatures is to separate oxygen and nitrogen from air and to liquefy them.

4.1.3 Refrigeration Equipment 制冷设备

制冷设备通常包含蒸发器、冷凝器、发生器等。

The main equipment components of the vapor compression refrigeration system are the familiar evaporator, compressor, and condenser. The equipment may be separate or of the unitary (also called self-contained) type. Unitary equipment is assembled in the factory. The household refrigerator is a common example of unitary equipment. Obvious advantages of unitary equipment are that it is more compact and less expensive to manufacture if made in large quantities.

There is a variety of commercial refrigeration equipment; each has a specific function. Reach-in cabinets, walk-in coolers, and display cases are widely used in the food service business. Automatic ice makers, drinking water coolers, and refrigerated vending machines are also commonly encountered equipment.

Air conditioning includes the heating, cooling, humidifying, dehumidifying, and cleaning (filtering) of air in internal environments. Occasionally it will be necessary to mention some aspects of air conditioning when we deal with the interface between the two subjects. A study of the fundamentals and equipment involved in air conditioning is nevertheless of great value even for those primarily interested in refrigeration.

4.2 The Reversed Carnot Cycle

逆卡诺循环

The Carnot cycle is a totally reversible cycle

that consists of two reversible isothermal and two isentropic processes. It has the maximum thermal efficiency for given temperature limits, and it serves as a standard against which actual power cycles are compared.

Since it is a reversible cycle, all four processes that comprise the Carnot cycle can be reversed. Reversing the cycle will also reverse the directions of any heat and work interactions. The result is a cycle that operates in the **counterclockwise** direction on a *T-s* diagram, which is called the reversed Carnot cycle. A refrigerator or heat pump that operates on the reversed Carnot cycle is called a Carnot refrigerator or a Carnot heat pump.

Consider a reversed Carnot cycle executed within the saturation dome of a refrigerant, as shown in Fig. 4.2. The refrigerant absorbs heat isothermally from a low-temperature source at T_L in the amount of Q_L (process 1-2), is compressed isentropically to state 3 (temperature rises to T_R), rejects heat isothermally to a high-temperature sink at T_R in the amount of Q_R (process 3-4), and expands isentropically to state 1 (temperature drops to T_L). The refrigerant changes from a saturated vapor state to a saturated liquid state in the condenser during process 3-4.

counterclockwise
逆时针的

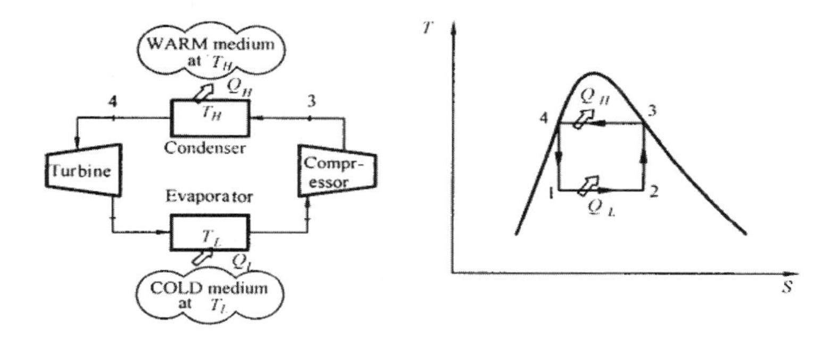

Fig. 4.2 The reversed Carnot cycle.

The coefficients of performance of Carnot refrigerators and heat pumps are expressed in terms of temperature as

$$COP_{R,\text{Carnot}} = \frac{1}{T_H / T_L - 1} \tag{4-1}$$

and

$$COP_{HP,\text{Carnot}} = \frac{1}{1 - T_L / T_H} \tag{4-2}$$

Notice that both COPs increase as the difference between the two temperatures decreases, that is, as T_L rises or T_R falls.

The reversed Carnot cycle is the most efficient refrigeration cycle operating between two specific temperature levels. Therefore, it is natural to look at it first as a prospective ideal cycle for refrigerators and heat pumps. If we could, we certainly would adapt it as the ideal cycle. As explained below, however, the reversed Carnot cycle is not a suitable model for refrigeration cycles.

The two isothermal heat transfer processes are not difficult to achieve in practice since maintaining a constant pressure automatically fixes the temperature of a two-phase mixture at the saturation value. Therefore, processes 1-2 and 3-4 can be approached closely in actual evaporators and condensers. However, processes 2-3 and 4-1 cannot be approximated closely in practice. This is because process 2-3 involves the compression of a liquid-vapor mixture, which requires a compressor that will handle two phases, and process 4-1 involves the expansions of **high-moisture-content** refrigerant.

It seems as if these problems could be eliminated by executing the reversed Carnot cycle outside the saturation region. But in this case we will have difficulty in maintaining isothermal conditions during the heat-absorption and heat-rejection processes. Therefore, we conclude that reversed Carnot cycle cannot be approximated in actual devices and is not a realistic model for refrigeration cycles. However, the reversed Carnot cycle can serve as a standard against which actual refrigeration cycles are compared.

4.3 Refrigerators and Heat Pumps

制冷机和热泵

high-moisture-content
高含水量的

We all know from experience that heat flows

in the direction of decreasing temperature, that is, from high-temperature regions to low-temperature ones. This heat-transfer process occurs in nature without requiring any devices. The reverse process, however, cannot occur by itself. The transfer of heat from a low-temperature region to a high-temperature one requires special devices called refrigerators.

制冷系数，也称制冷性能系数，是指单位功耗所能获得的冷量，是制冷系统的一项重要技术经济指标。

Refrigerators are cyclic devices, and the working fluids used in the refrigeration cycles are called refrigerants. A refrigerator is shown schematically in Fig.4.3(a). Here Q_L is the magnitude of the heat removed from the refrigerated space at temperature T_L, Q_H is the magnitude of the heat rejected to the warm space at temperature T_H, and $W_{\text{net,in}}$ is the net work input to the refrigerator. Q_L and Q_H represent magnitudes and thus are positive quantities.

热泵是一种充分利用低品位热能的高效节能装置。

Another device that transfers heat from a low-temperature medium to a high-temperature one is the heat pump. Refrigerators and heat pumps are essentially the same devices; the difference is in their objectives only. The objective of a refrigerator is to maintain the refrigerated space at a low temperature by removing heat from in. Discharging this heat to a higher-temperature medium is merely a necessary part of the operation, not the purpose. The objective of a heat pump, however, is to maintain a heated space at a high temperature. This is accomplished by absorbing heat from a low-temperature source, such as well water or cold outside air in winter, and supplying this heat to a warmer medium such as a house (Fig.4.3[b]).

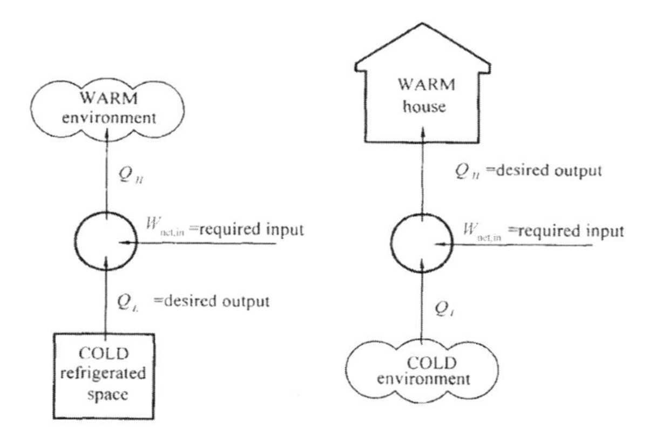

(a) refrigerator (b) heat pump

Fig. 4.3 Heat transfers from low temperature regions to high temperature regions.

The performance of refrigerators and heat pumps is expressed in terms of the coefficient of performance (COP), which was defined as

$$COP_R = \frac{\text{Desired output}}{\text{Required input}} = \frac{\text{Cooling effect}}{\text{Work input}} = \frac{Q_L}{W_{\text{net,in}}} \quad (4\text{-}3)$$

$$COP_{HP} = \frac{\text{Desired output}}{\text{Required input}} = \frac{\text{Heating effect}}{\text{Work input}} = \frac{Q_H}{W_{\text{net,in}}} \quad (4\text{-}4)$$

These relations can also be expressed in the rate form by replacing the quantities Q_L, Q_H, and $W_{\text{net,in}}$ by $\dot{Q}_L$, $\dot{Q}_H$ and $\dot{W}_{\text{net,in}}$ respectively. Notice that both COP_R and COP_{HP} can be greater than 1.

A comparison of Eqs. (4-3) and (4-4) reveals that

$$COP_{HP} = COP_R + 1 \qquad (4\text{-}5)$$

for fixed values of Q_L and Q_H. This relation implies that $COP_{HP} > 1$ since COP_R is a positive quantity. That is, a heat pump will function, at worst, as a resistance heater, supplying as much energy to the house as it consumes. In reality, however, part of Q_H is lost to the outside air through piping and other devices, and COP_{HP} may drop below unity when the outside air temperature is too low. When this happens, the system normally switches to the fuel (natural gas, propane, oil, etc.) or resistance-heating mode.

The cooling capacity of a refrigeration system—that is, the rate of heat removal from the refrigerated space—is often expressed in terms of tons of refrigeration. The capacity of a refrigeration system that can freeze 1 ton of liquid water at 0 °C into ice at 0 °C in 24 h is said to be 1 ton. One ton of refrigeration is equivalent to 211 kJ/min. The cooling load of a typical 200 m^2 residence is in the 3 ton (10 kW) range.

Fig.4.4 illustrates some applications in a variety of fields and industries employing refrigeration.

Fig. 4.4 Applications of refrigeration in engineering.

4.4 Refrigerants
制冷剂

制冷剂，又称冷媒、致冷剂、雪种，是各种热机中借以完成能量转化的媒介物质。

The comparative performance of refrigerants commonly used in the vapor compression cycle is given in Tab. 4.1, the values of which were all obtained by calculation, using the methods of this chapter. Actual operating conditions will therefore be somewhat different, owing to the effects of the factors.

In general, it has been assumed that the vapor enters the compressor in a saturated condition at 5 °C. In refrigerants 113 and 114, however, where saturated suction gas would result in condensation during compression, enough superheat has been assumed to ensure saturated discharge gas. This superheat has not been counted as part of the refrigerating effect.

A condensing temperature of 40 °C has been taken for all refrigerants.

The performances have been calculated on the assumption that compression is isentropic.

In reciprocating compressors, the displacement (L · [s · kW $^{-1}$]) should be low so that the required performance can be achieved with a small machine. For a centrifugal compressor, on the other hand, large displacements are desirable in order to permit the use of large gas passage.

The reduced frictional resistance to such passages improves the compressor efficiency.

Tab. 4.1 Comparative performance of refrigerants evaporating at 5 °C and condensing at 40 °C

Refrigerant number	Suction temp °C	Evaporating pressure bar	Condensing pressure bar	Compression ratio
718	5	0.009	0.074	8.46
11	5	0.496	1.747	3.52
717	5	5.160	15.55	3.01
114	12.7	1.062	3.373	3.18
12	5	3.626	9.607	2.65
113	10.4	0.188	0.783	4.16
134a	5	3.451	10.032	2.91
22	5	5.838	15.34	2.63
502	5	6.678	16.77	2.51
Refrigerating effect $kJ \cdot kg^{-1}$	**Specific vol. of vapor $m^2 \cdot kg^{-1}$**	**Compressor displacement $L \cdot (s \cdot kW)^{-1}$**	**Power in kW per kW refrigeration**	**Carnot cycle efficiency %**
2370.0	147.0	62.0	0.1355	92.9
157.0	0.332	2.12	0.1395	90.2
1088.0	0.243	0.214	0.1456	86.4
106.2	0.122	1.14	0.1484	84.8
115.0	0.047	0.109	0.1502	83.8
129.5	0.652	5.03	0.1511	83.3
145.0	0.0584	0.403	0.1516	83.0
157.8	0.040	0.255	0.1518	82.9
101.0	0.026	0.259	0.1632	77.1

*718 Water, H_2O

11 Trichlorofluoromethane, CCl_3F

717 Ammonia, NH_3

114 Dichlorotetrafluoroethane, $CClF_2$-$CClF_2$

12 Dichlorodifluoroethane, CCl_2F_2

113 Trichlorotrifluoroethane, CCl_2FCClF_2

134a Tetrafluoroethane, CF_3CH_2F

22 Chorodifluoromethane, $CHClF_2$

502 Azeotropic mixture (48.8% R22, 51.2% R115) for low temperature work

Note. ① An azeotrope is a mixture of refrigerants, the liquid and vapor of which, in equilibrium, have a constant boiling point.

② Bar is the abndoned unit of pressure. 1 bar=100,000 Pa.

The power required to drive the compressor is obviously very important because it affects both the first cost and the running cost of the refrigeration plant.

Although not listed in the table, the critical and freezing temperatures of a substance have also to be noted when assessing its suitability for use as a refrigerant.

There are, of course, factors other than thermodynamic ones which have to be taken into consideration when choosing a refrigerant for a particular vapor compression cycle. These include the heat transfer characteristics, dielectric strength of the vapor, inflammability, toxicity, chemical reaction with metals, tendency to leak and leak detectability, behavior when in contact with oil, availability and, of course, the cost.

The general concern about the ozone depletion and global warming effects of the refrigerants commonly in use has stimulated searching for alternative, more acceptable refrigerants. Among those considered are: well-established refrigerants such as **ammonia** (provided proper safety measures are adopted), **hydrocarbons** (also requiring safety precautions) and mixtures of two or more refrigerants, which introduce unusual effects and are incompatible with conventional mineral oils for **lubrication** of the compressor.

ammonia
氨

hydrocarbons
碳氢化合物

lubrication
润滑

Butler discussed refrigerant mixtures with components that boil at different temperatures, giving changes in the boiling points (termed "glide") as the mixture evaporates or condenses. "Bubble point" is the temperature at which a refrigerant liquid just starts to evaporate and "dew point" is the temperature at which the vapor starts to condense. With a single refrigerant these points coincide and a zeotropic mixture having liquid and vapor phases in equilibrium also has a constant boiling point. On the other hand, for a zeotropic mixture of refrigerants, the constituents have different boiling points and the difference is termed the glide value for the mixture. Glide may lead to differential frosting temperatures across an evaporator and condensers and evaporators tend to be larger than for a single refrigerant. Changes in heat transfer characteristics and refrigerant handling difficulties are possible with glide. Note that a refrigerant and a lubricant are a zeotropic mixture requiring special consideration. Whereas hydrocarbon refrigerants are compatible with conventional mineral oils, HFC refrigerants (such as R134a) and mixtures of them are not and a polyester lubricant is required.

4.4.1 Ozone Depletion Effects
臭氧破坏效应

The refrigerants that have been widely used in

air conditioning and other applications throughout the world since the nineteen thirties, comprise molecular combinations of chlorine, fluorine, carbon and hydrogen. Reference to the footnotes in Tab.4.1 verifies this. There is a significant turnover in the quantity of refrigerant used for topping-up purposes in plant, as leaks occur and maintenance and replacement work is carried out. Butler has suggested that this amounts to 80 percent of total usage of the refrigerant for air conditioning purposes.

臭氧层损耗：在大气层的平流层内离地面20 ~ 30 km的地方是臭氧的集中层带。臭氧层是地球最好的保护伞，它吸收了来自太阳的大部分紫外线。然而，近20年的科学研究和大气观测发现，每年春季南极大气中的臭氧层一直在变薄，事实上在极地大气层中存在一个臭氧“洞”。

Fully halogenated refrigerants are termed chlorofluorocarbons (CFC). They are chemically very stable and exist for a long time in the atmosphere, after leaking from a refrigeration plant. In due course they rise into the stratosphere and, at a height of between six and thirty miles above the surface of the earth, the molecules break down in the presence of solar radiation, losing an atom of chlorine.

Two reactions then take place. Firstly, the chlorine combines with a molecule of ozone to form chlorine monoxide and oxygen:

$$Cl+O_3 \rightarrow ClO+O_2 \tag{4-6}$$

Secondly, the molecules of chlorine monoxide then combine with oxygen to form oxygen molecules and chlorine molecules.

$$2ClO+O_2 \rightarrow 2Cl+2O_2 \tag{4-7}$$

The chlorine molecules are once again free to combine with ozone and the process repeats up to 10,000 times, some CFCs having an atmospheric half life of the order of 100 years. The net result appears to be a steady depletion of the ozone content of the atmosphere at high altitudes. Since ozone is the agent in the upper atmosphere that prevents the entry of excessive, dangerous, and ultraviolet radiation, the consequences of ozone depletion are obvious and serious.

An ozone depletion factor (ODF) has been calculated for each refrigerant, in relation to value of 1.0, assigned to R11, which is the worst offender because it contains the largest proportion of chlorine by weight.

If one of the chlorine atoms in the molecule of a CFC refrigerant is replaced by a hydrogen atom, the stability of the molecule is much reduced and its ozone depleting effect becomes smaller. This has encouraged the use of such refrigerants, which are termed hydrochlorofluorocarbons (HCFC), as an interim measure, before the use of CFC refrigerants can be fully phased out. An example of an HCFC refrigerant is R22, which has been extensively used for many years in reciprocating compressors, but is no longer acceptable.

Refrigerants with no chlorine present have

ODF values of zero. Such refrigerants are termed hydrofluorocarbons (HFC).

Of the refrigerants listed, R134a offers promise, being similar in many respects to R12 but with no ozone depletion effect.

4.4.2 Global Warming 全球变暖

The British Refrigeration Association Specification describes a way of comparing the influence of different refrigeration systems on climate change. A value, termed the Total Equivalent Warming Impact (TEWI) is determined. This estimates a direct effect, which is the total release of refrigerant and an indirect effect, which is an estimate of the total energy consumption, both expressed in equivalent tons of CO_2 over the expected life of the system. The sum of the two effects is the TEWI value. Apparently, the indirect effect of energy consumption usually dwarfs the direct effect of refrigerant release. Miro and Cox have pointed out that HFC refrigerants, such as R134a, have a much greater global warming potential on a molecule-to-molecule basis than CO_2 but their overall contribution is trivially small when the enormous abundance of CO_2 is considered. HFCs contain no chlorine. See Tab. 4.2.

全球气候变暖是一种和自然有关的现象，是由于温室效应不断积累，导致地气系统吸收与发射的能量不平衡，能量不断在地气系统累积，从而导致温度上升，造成全球气候变暖。

Tab. 4.2 Ozone depletion factors

Refrigerant	ODF	Proportion of chlorine by weight/%	Atmospheric half life in years
CFCs			
R11	1.0	77.40	65
R12	1.0	58.6	146
R113	0.8	56.7	90
R114	0.8	41.5	185
R115	0.4	22.9	380
HCFCs			
R22	0.05	41.0	20
R142b	0.06	35.2	13
R141b	0.1	60.6	10
R125	0.02	26.0	4
R123	0.02	46.3	2
HFCs			
R152a	0	0	1
R134a	0	0	6

4.5 Other Methods of Refrigeration

其他形式的制冷

制冷剂应满足不易燃易爆，无毒或低毒，无刺激性气味，对人体的健康无损害等要求。

There are several other forms of refrigeration, the most important of which is vapor absorption. Other methods, considered by Butler and Perry, are as follows.

(1) Water vapor systems. The vapor pressures to achieve effective cooling are very low, making the method impractical. With a steam jet as the source of energy the COP is less than 0.3.

(2) Stirling cycle. In principle, this should have a higher COP than conventional vapor compression. However, hot and cold heat transfer areas are very small, which introduce difficulties. The method has been used in small domestic refrigerators.

(3) Acoustic cooling. Alternate compressions and expansions, causing heating and cooling, are produced in a closed tube by a sound generator. Efficiencies appear to be poor and the comparatively large.

(4) Magnetic cooling. Some metals heat up when magnetized and cool when demagnetized. A prototype plant at the US Department of Energy Ames Laboratory, in Iowa, has rotated a metal disc continuously through strong and weak magnetic fields with heat transferred by water or an anti-freeze fluid to heat exchangers. Initial efficiencies are claimed to be 20 percent better than conventional methods but with higher capital costs and a simple pay back of 5 years.

(5) Pulse tube cooling. This is similar to acoustic cooling but with a compressor instead of a sound generator, and an inert gas in the tube.

(6) Thermo-electric cooling. This makes use of the Peltier effect and is commercially available with cooling capacities up to 500 W. It has a high capital cost and very low COP.

(7) Thermionic cooling. Cooling occurs at the negative electrode when high energy electrons flow between two electrodes in a vacuum. A better COP than vapor compression gives is possible if semiconductor materials are used. It is potentially cheap with future development.

(8) Air cycle cooling. This is well established for aircraft, but with high capital cost and low efficiency. COPs are in the range of 0.35 to 0.57. Air is compressed and expanded without a phase change of refrigerant.

4.6 Safety
安全性

Whichever refrigerant is used it must be safe. This is dealt within BS 4434:1989, covering the design, construction, and installation of refrigeration plant and systems. Refrigerants are classified in three groups:

(1) These are non-inflammable in vapor form at any concentration in air at standard atmospheric pressure and 20 °C. They have a low toxicity although when in contact with a flame or a hot surface toxic products of decomposition may form. Examples are: R11, R12, R13, R22, R113.

(2) Toxicity is the dominant feature with these refrigerants and it is almost impossible to avoid

a toxic concentration if an escape of refrigerant occurs. An example is: R717 (ammonia).

(3) These are inflammable and are an explosive hazard, although with a low order of toxicity. Examples are: R170 (ethane), R290 (propane), R600 (butane).

Group 3 refrigerants should not be used for institutional or residential buildings, or those buildings used for public assembly.

本章重点内容介绍

古代文明在自然条件下使用制冷。罗马统治者让奴隶从高山上运送冰雪，用来保存食物并且在天气炎热时提供冷饮。今天，制冷行业是社会的重要组成部分。 制冷的应用分为以下几类：家用，商用，工业，空调。制冷，通常被称为冷却过程，更正确的定义为用人为的方法在一定的时间或空间内将物体或流体冷却，使其温度降到环境温度以下并维持这一低温。机械制冷最普遍的方法是蒸气压缩系统。另一种广泛使用的方法为吸收式制冷系统。其他制冷方法有热电制冷、蒸气喷射式制冷、空气循环制冷。蒸气压缩制冷系统的主要设备部件是我们熟悉的蒸发器、压缩机、冷凝器和节流装置，每种设备都有特定的功能。

理想的制冷循环——卡诺循环是一个完全可逆的循环，由两个可逆的等温过程和两个等熵过程组成。它在给定的温度限制下具有最大的热效率，并且它经常用来作为评价实际循环的标准。

我们从经验中知道，热量是沿着温度降低的方向流动的，也就是说，热量从高温传递到低温。反向过程不能自发进行。制冷机就是从低温热源吸热，向高温热源放热，从而达到制冷的效果。另一种将热量从低温介质转移到高温介质的装置是热泵。制冷机和热泵本质上是相同的设备，区别在于它们的目的不同。制冷机的目的是通过去除内部的热量来保持冷藏空间处于低温。而热泵的目的是在高温下保持一个加热的空间。

通过制冷剂的热力性质表来判断对于不同循环，应该应用哪个种类的制冷剂。制冷剂广泛应用于空调和其他领域，它是包括氯、氟、碳和氢的分子组合。因此，在使用的过程中就容易导致臭氧层被破坏，从而使全球气候变暖。必须注意制冷剂的安全性，即无论使用哪种制冷剂，它都必须是安全的。

Chapter 5
Floor Heating
地板采暖

5.1 Introduction
简　介

Floor heating system, is short for **radiant floor heating system**. Radiant heat dates to ancient times—the Romans used radiant floor heating in their bathhouses. And for centuries, the Koreans heated their royal palaces and traditional homes in this manner. Today, radiant floor heating technology has been improved and can be used in all or part of our homes.

地热采暖全称为低温地板辐射采暖，简称地暖，是以不高于 60 ℃ 的热水为热媒，在加热管内循环流动，加热地板，通过地面以辐射和对流的传导方式向室内供热的供暖方式。

Radiant floor heating is a method of heating our homes by applying heat underneath or within the floor. In other words, it turns a floor into a large-area, low-temperature radiator. Comparable to warming yourself in the sun, this type of heating warms objects as opposed to raising the temperature of the air.

radiant floor heating system 地暖辐射采暖系统

There are three types of radiant floor heating:

hydronic, electric and air. Hydronic (liquid) system is the most popular and cost-effective radiant floor heating system now. The following focuses on hydronic (water) radiant floor heating system, which is often called low temperature hot water radiant floor heating system, which pumps heated water (generally below 60 ℃) from a boiler through tubing laid in a pattern underneath the floor, and then the cooler water returns to the heat source where it is reheated and sent out again in what is known as a "closed-loop system". In the case, the thermal mass of the slab holds heat and radiates it slowly to the living space above, so the temperature throughout the room remains nearly constant. It is said that this factor alone reduces heat loss by up to 25 percent when compared to similar homes using conventional heating methods.

For hydronic radiant floor heating system, metal (like copper) tubes/pipes have been used in the past, but most systems today use plastic tubes/pipes (like rubber or cross-linked **polyethylene** [PEX]) — the latter being by far the most common. These pipes form a continuous loop between two central manifolds. Each room has its own circuit of pipes and can be controlled regulating the flow of hot water putting the heat exactly where you want it (Fig.5.1).

hydronic
循环

polyethylene
聚乙烯

Fig. 5.1 PEX tubing has been secured in place before pouring a gypsum concrete slab.

5.2 System Components
系统组成

There are three components to hydronic radiant floor heating system: a heat source, a **distribution piping system** and controls (Fig.5.2).

The heat source in this heating system is usually a boiler or a hot water heater, but other heat sources can be used too. The energy used to heat the hot water can be natural gas, oil, electricity, propane, wood or solar hot water collection.

A **circulator pump** (a specific type of pump used to circulate gases, liquids, or slurries in a closed circuit. They are commonly found circulating water in a hydronic heating or cooling system. Because they only circulate liquid within a closed circuit, they only need to overcome the friction of a piping system [as opposed to lifting a fluid from a point of lower potential energy to a point of higher potential energy]) near the water

distribution piping system
配水管道系统

circulator pump
循环泵

supply manifold moves the water from the mixing valve to the supply manifold into the distribution piping system (tubing) inside the floors.

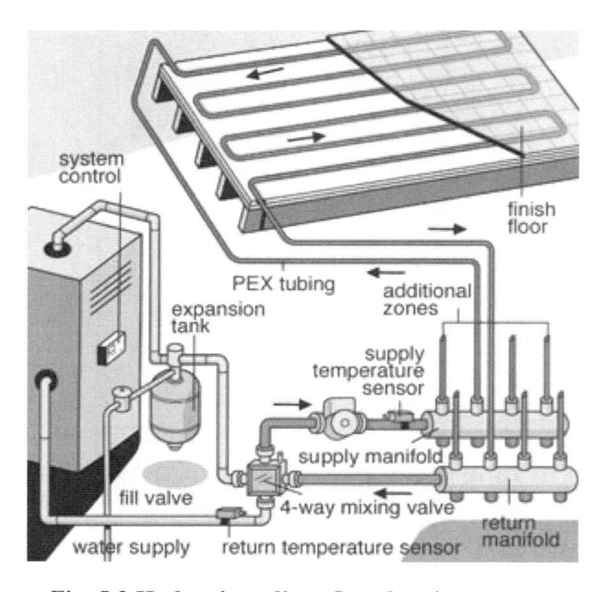

Fig. 5.2 Hydronic radiant floor heating system.

To select how warm or cool a room or home will be, controls are required to set the system to a particular temperature. A manifold system with **thermostat** or **aquastat switches** typically located in an accessible wall cavity and provides a series of simple valves that are used to regulate the flow of water through each zone. There is a caution not to exceed the recommended maximum temperature as it could warp solid hardwood flooring and cause stress to the system.

aquastat switches
恒温器开关

thermostat
恒温器

5.3 Benefits
优 势

Floor heating system offers a number of significant benefits:

地热采暖的优势：散热均匀，舒适，清洁健康，环保节能，美观大方，高效隔音。

(1) Comfort. By far, maybe the biggest selling point for radiant floor heating is comfort. The large radiant surface means that most of the heat will be evenly distributed and delivered by radiation rather than by convection (the primary mechanism of heat delivery from conventional hydronic baseboard "radiators"). Warmer surfaces in a living space result in a higher mean radiant temperature. With higher mean radiant temperatures, most people are comfortable even at lower air temperatures.

(2) Energy Saving. Proponents of radiant floor heating argue that someone normally comfortable at 22 °C will be comfortable in a building with radiant floor heating kept at 20 °C. And this mean radiant floor heating is more economical to operate at 20 °C rather than the usual 22 °C. So people could keep their thermostats lower and thus realize significant energy savings. Besides this, zoning a variety of rooms with the options for different temperatures has the potential to reduce energy consumption.

(3) Energy Source Compatibility. Since radiant

floor heating has a low operating temperature, it is feasible for alternative heat sources such as solar panels, a heat pump, and a heat recovery system. These systems can be used in any combination to supplement the output from your boiler and save you even more of your fuel costs.

(4) Increased Boiler Life. By operating a boiler at a lower temperature, its life can be extended. Radiant floor heating system typically uses water temperatures of 30~60 ℃, compared with baseboard hydronic systems that typically operate at 55~70 ℃. At these operating temperatures, boiler life can exceed 45 years, according to information from DOE (Department of Energy).

(5) Quietness. Radiant floor heating hydronic system is extremely quiet. Unlike forced-air heat, there is no noise from a fan or airflow through ducts; and unlike hydronic baseboard heat, there is usually no gurgle of water through baseboard radiators or creaking from expansion and contraction. The primary noise will be the sound of circulating pumps and the fan used in power-venting the boiler.

(6) Cleanliness. An argument can be made for improved indoor air quality in houses with radiant floor heating. Unlike conventional forced-air furnaces, radiant floor heating has no ducts or radiators to contribute to dust collection or movement. And unlike electric baseboard,

there will be no surfaces hot enough to burn dust particles — which could introduce volatile chemicals or toxic particulates into house air (even passing through filters). This concern would be the greatest for people with acute chemical sensitivities.

(7) Flexible Room Layout. Radiant floor heating systems are "invisible"; in other words, there are no baseboard radiators or air registers with radiant floor heating. This means you can put your furniture wherever you like. Bathrooms or special use areas with hard floor finishes are well suited to this type of heating.

5.4 Design and Installation
设计与安装

Prior to the installation of a system, a qualified floor heating specialist should make a heating-load estimate of your home on a room-by-room basis. The heating-load estimate will assist in an efficient system design. By placing the tubing in specific patterns and spacings, the system can accommodate the insulation of the room/home and flooring choices.

It should be noted that if any later flooring renovation is undertaken, the hydronic radiant floor heating installer should be notified to make any required adjustments to the heating system.

For example, the water temperature of the heating system would need to be adjusted if there was a change from a bare or painted finished floor slab to ceramic tile, or wood flooring or to carpet with underlay. Wood flooring and thick carpets act as an insulation blanket, restricting upward heat flow and reduce the efficiency of the system.

Here introducing two choices of installation:

(1) **Slab-on-grade system**

One example of a slab-on-grade system is PEX tubing attached to a wire mesh or clipped onto rigid Styrofoam insulation. Concrete is poured over the piping or tubing at the ground "grade" level. The slab must be insulated from the exterior side of the floor all the way to the slab edges.

(2) **Thin slab system**

① The floor heating tubing is fastened above the subfloor and is covered with lightweight concrete or selflevelling gypsum cement underlayment. The floor ranges in thickness from 3.2 to 3.8 cm.

② Another version is to sandwich the tubing between the subfloor and the finished floor. This raises the floor only 1.3 cm. There are a variety of new underlayment panels that hold the tubing in place and incorporate aluminum transfer plates to

slab-on-grade system
地板系统

thin slab system
薄板系统

improve heating performance.

It is recommended that a licensed contractor install the heating system.

5.5 Costs

经济性

An approximate cost of an installed hydronic radiant floor heating system by a licensed mechanical contractor can range from $60 to $80 approximately per square meter. This cost can be more or less depending on specific heating requirements and energy efficiency results. In addition to the heating system, a mechanical ventilation system is required in the house.

5.6 Cautions and Solutions

注意事项与解决方法

Due to thermal mass, the system may be slower to respond to temperature changes. Overheating can occur in poorly controlled or zoned systems. The system is not designed to have the temperature frequently adjusted.

This is not a do-it-yourself project. It requires professional installation, maintenance and repair. Having professionals do the installation will allow you to have the best performance and warranty on the heating system.

5.7 Conclusions
结　论

On the whole, radiant floor heating offers the most efficient and comfortable living conditions in any climate, including unsightly base boards, reducing heat loss, smoothing temperature transition, no drafts, and creating a quieter home. Also design flexibility, ease of installation, reliable components, make floor heating the ideal choice for any home.

In a word, it combines radiation, conduction and convection to create the ideal thermal environment for health and comfort. So it is believed that more and more homeowners and builders will choose radiant heat, and the application of floor heating system will have a far-reaching foreground.

本章重点内容介绍

关于地暖系统，它是地板辐射采暖的简称，它的历史可以追溯到古罗马时期。古罗马人将地暖用于浴室，到了现代地暖技术已经步入我们的日常生活中了。地暖是一种使物体升温，通过辐射来使得室内温度升高的技术。地暖有三种形式：循环式，电热式和空气式。循环式是目前用得比较多的地暖形式，它是将管道铺设在地板下面，使得热水在管道中循环，是非常经济实用的。目前大部分水地暖系统用的都是塑料管，每个房间都有自己的回路并且可以控制水温。地暖由热源、管道和控制系统三部分组成。大部分的热源是由锅炉或热水器供热，也有使用其他能源的。循环泵的作用是使得供给的热水在管道中能循环流动，通过使用恒温器或带有恒温器开关的歧管系统并且和阀门相连接，用于调节流经每个区域的水流量。供水的温度不能很高，否则它会使得地板翘曲。

在安装地暖前，专业人员都会计算出每个房间所需的热负荷，并结合现有图纸设计出一个合理的管道图。如果以后地板翻新，要根据铺在管道上的材料重新设定水温。

成本可能或多或少取决于具体的供热要求和能效结果。除了供暖系统外，还需要在室内安装机械通风系统。系统控制不良容易发生局部过热。该系统需要专业人士进行安装、维护和修理。

总的来说，地板辐射供暖在任何气候条件下都能提供最高效、最舒适的居住条件，同时设计灵活，安装方便，元件可靠，地暖已成为任何家庭的理想选择。

Chapter 6
Ventilation
通　风

6.1 Introduction
简　介

通风系统是借助换气稀释或通风排除等手段，控制空气污染物的传播与危害，实现室内外空气环境质量保障的一种建筑环境控制技术。通风系统通常包括进风口、排风口、送风管道、风机、过滤器、控制系统以及其他附属设备等装置。

All occupied spaces need **ventilation**, to maintain good air quality and a comfortable temperature. Which of these constraints is the most important varies across the world, and with the time of year. Air quality is reduced by **respiration**, as the CO_2 percentage increases, by the build-up of chemicals given off from the building fabric and furnishings, and by smells, for example from cooking. **Incidental** heat input to building comes from people (over 100 W each!), electrical machinery and solar gains, as well as any heating we may intentionally provide. In the UK in summer, the removal of heat usually requires a higher flow rate (in m^3/s, or comparable units) than maintenance of air quality.

ventilation
通风，通风设备

respiration
呼吸，呼吸作用

incidental
附带的，偶然的

One solution to the provision of ventilation is a

mechanical system, either just using fans at inlets and/or extraction points to force a flow through the space, or fans coupled to an air handling unit which also controls the temperature and humidity of the incoming air (air conditioning). If well designed, maintained and run, these systems can create a precisely controlled comfortable environment, but they are very expensive and **energy-extensive** (up to half the energy use in an air-conditioned building can be taken up by these systems), and introduce other problems. Without cleaning, bacteria can build up in the ducts and cause health problems, the noise can be intrusive, and the lack of control over the environment can lead to occupant dissatisfaction.

Of course, buildings without mechanical systems are still ventilated, but by a kind of air movement now called natural ventilation. In this kind of ventilation, air movements occur because of pressure differences created by natural effects as shown in Fig.6.1, either dynamic pressure differences due to the flow of the wind around a building, or static pressure differences due to the build-up of warm air within a building (the stack pressure). These are free! However they are not so easy to control as an air-conditioning system, so we are working to understand more about natural ventilation flows, to help in the design of more comfortable buildings. Traditional buildings tend to ventilate well because of design constraints and long-established rules of thumb — the walls are

energy-extensive
高能耗的

quite solid (giving a high thermal mass), while the windows and the floor - plan are small (reducing solar gains and the area/perimeter ratio). However with the introduction of steel, concrete and strong glass, we can now build buildings that look completely different (like those on the Sidgwick Site at the University of Cambridge shown in Fig.6.2 and Fig.6.3), and that introduce new challenges for ventilation.

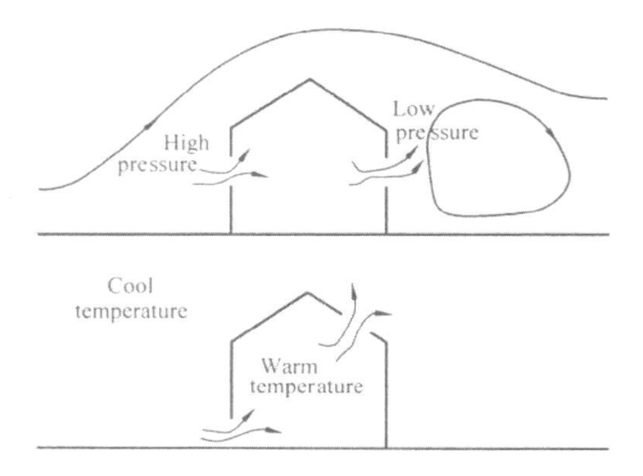

Fig. 6.1 Natural ventilation.

Fig. 6.2 History Faculty, CU.

Fig. 6.3 Law Faculty, CU.

6.2 Application of Natural Ventilation 自然通风的应用

Natural air change takes place in buildings as a result of wind and the difference in density between the indoor and outdoor air. Without control, such natural infiltration is **haphazard**, and the process can only legitimately be termed "ventilation" if the **arrangements** are designed to maintain the desired state of the indoor air under a variety of outdoor conditions.

Temperature differences and wind speed can cause the transfer of enormous quantities of air. For instance, measurements show that the natural air change in an open-hearth plant or a rolling mill amounts to about 20 million kg/h. In forges, ironworks and other hot shops the air transfers may also be millions of kilograms per hour.

A very large consumption of energy would be required to move such quantities by mechanical means. The great economic importance of natural ventilation is that it can bring about these air changes without expenditure of mechanical energy.

The time has long passed when it was necessary to **demonstrate** the benefits of natural ventilation and justify its application. The proofs were simple and very convincing. They were based on comparisons of mechanical ventilation

haphazard
偶然的，随意的

arrangement
布置，整理，安排

demonstrate
演示，示范

and natural ventilation. In hot shops where all the **emphasis** was laid on mechanical ventilation, and natural air change, being regarded as unimportant, was not taken into account at all, it was found in all the tests that the volume of natural air change many times exceeded the volume of mechanical ventilation. This revealed the **negligible** role of general mechanical ventilation despite its heavy installation and running costs. Mechanical ventilation in these cases was best used as a corrective to natural ventilation, in the form of air curtains and local air supply or extraction.

In the hot season of the year natural ventilation can be used in almost every branch of industry, except the comparatively few industrial undertakings which require per-treatment of the air for technological reasons. In single-span workshops the outdoor air enters the premises through vents at the foot of the walls, and vitiated air is removed from the shop through louvers in the roof. In multi-span shops, single vents in the walls of the building are not enough, and for the ventilation of working spaces far from the walls additional air has to be introduced from the roof via the spaces between the roof bays. In this arrangement it is necessary to have a good air supply through these spaces and to arrange hot and cold spans alternatively. The outdoor air enters the building through the openings in the roof of the cold spans and is then distributed to the adjoining hot spans. The use of natural

emphasis
重点，强调

negligible
微不足道的
可以忽略的

ventilation in winter assumes that the indoor excess of production of heat will be sufficient to heat the estimated volume of air, and that besides the openings in the roof bays, there will be a series of vents (5~7 m above floor level) for the entry of cold air. The height of these vents is so calculated as to ensure that the incoming air is heated by mixing with the internal warm air while falling from this height to the work area.

Natural air change is above all suitable workshops in which a great deal of heat is liberated ("hot shops"), viz. blast furnaces, open-hearth furnaces, rolling mills, forges, foundries, and heat-treatment shops, boiler houses, engine rooms and so on. A combination of general natural and local mechanical ventilation, as already mentioned, is often specially useful in these cases.

The effectiveness of natural ventilation depends on many factors which must be considered in using the building as well as in planning it.

The main factors on which successful control of natural ventilation of industrial premises depend are the layout and sitting of the sources of heat, the design of the building (number of spans, form and shape of roof) and the arrangement of ventilation openings in walls and roof bays. The most satisfactory solutions are obtained when the architect and the engineer collaborate and take account of problems of natural ventilation as well as those numerous and complex requirements that industrial undertakings have to satisfy at the design stage.

Many industrial buildings exist in which natural ventilation was provided for in the design stage and has since been satisfactory in use. On the other hand, there are numerous instances in which natural ventilation did not receive sufficient attention in the design of the premises. This lack of attention is now being paid for very dearly.

To clarify the problems of natural ventilation we shall first of all discuss the effects of heat sources in a building in relation to the air currents they cause in single-bay and multi-bay buildings. The effect of wind on a building and the resulting pressure distribution will also be considered.

In view of the great complexity of a theoretical study of these phenomena, they are best investigated experimentally with the aid of two- and three-dimensional models using air or water as the working medium.

6.3 Fundamentals of Industrial Ventilation 工业通风原理

The purpose of ventilation is to maintain in the building a prescribed condition and cleanliness of the air (in other words, the temperature, air velocity and concentrations). This task in the last analysis is resolved as follows. The vitiated air is removed from the building (extract ventilation), whilst in its place clean air is introduced, often specially treated (inflow ventilation).

In essence this boils down to heat transfer and mass transfer between the incoming air and the air already within the building. If owing to excessive internal heat production the temperature of the air in the building tends to exceed the specified norms, cooler air is introduced and mixed with the indoor air; the temperature of the air (owing to heat transfer) then remains at the norm. If harmful gases or vapors are released, their concentrations are held within specified limits by dilution with the clean incoming air.

More often than not mass and heat transfer take place simultaneously. For instance, the production of convective heat is very often accompanied by the releases of gases and highly dispersed dust.

Ventilation can be affected by fans (mechanical ventilation) or by the difference between the densities of the columns of internal and external air, and also by the action of wind (natural ventilation).

Ventilation can be general or local. Local extract ventilation is intended for removing polluted air at source, to prevent the dispersal of impurities throughout the building. As much of the impurity as possible is removed in this way so that a minimum has to be diluted by the incoming air. Local **exhaust** is not essentially ventilation proper.

Local ventilation thus limits the area of dispersal. This is assisted by the use of fixed screens or by air curtains. The impurity is removed by suction of the polluted air, and this can be combined with a jet of air which impels the impurity towards the suction opening.

If air is introduced into a building, some excess pressure is set up in it. In the steady state this pressure will be such that the local quantity of air leaving the building through specially provided vents, or through random cracks in the external surfaces, is equal to that which is introduced. A similar phenomenon will occur with the extract of air from the building. Here a negative pressure (rarefaction) is set up in the building, and in consequence air will be sucked in through gaps from outside and from **adjacent** rooms to take the place of the extracted.

In certain cases this air has an unfavorable effect. For instance, if cold outdoor air enters a building in which much water vapor is produced it would create mist on mixing with the internal hot and moist air. If the inflow from outside or from adjacent rooms satisfies the hygienic requirements, it can be used to replace general

exhaust
废气，排气

adjacent
相邻的，邻近的

mechanical ventilation by natural ventilation.

Ventilation is essentially the science of the control of air change in buildings.

In solving the problems of ventilation, the following questions arise:

(1) What quantity of air should be supplied to the building per unit of time, what quantity should be extracted and how?

(2) What characteristics should the incoming air have, and is preliminary treatment of the air necessary (heating, cooling, dehumidifying, conditioning, dust removal etc.)?

(3) What should be the disposition of the inlets and outlets?

(4) What should be the design of all the elements which determine the rate of air change?

To resolve the issues of general ventilation it is necessary to know the quantity of impurity entering per unit time into the air of the building. It is also essential to know how the impurity is dispersed within the building, and how its distribution can be influenced by ventilation.

By extracting the air from areas with high concentrations of impurity, one considerably reduces the quantity of air needed for ventilation.

For instance, in iron foundries the concentration of carbon monoxide (CO) in upper levels can be 0.04 g/m^3, whereas in the work area it should not exceed the permissible norm 0.02 g/m^3. This stratification of the concentration is maintained by a supply of fresh air near the ceiling, in descending it would disturb the stratification and mix with the vitiated air, and with the same air change the concentration of CO in the work area would be 0.03 g/m^3. To obtain a concentration of 0.02 g/m^3 one would have to increase the quantity of ventilation air by a factor of about 1.5. Thus the question of the estimated quantity of ventilating air is directly related to the question of the arrangements for ventilation.

To calculate and design local ventilation in the form of air douches, it is necessary to know the properties of the jet, the laws governing the variation of its velocity, temperature and concentration and the geometric dimensions of the jet. To obtain the hygienically prescribed parameters of the air at the workplace, one needs to know the initial parameters of the air and then find the forms of nozzles to produce a jet which would satisfy these requirements.

6.4 Fans

风 机

Air movement in warm air heating, ventilation

and air conditioning systems is made possible by the use of fans. A fan creates air movement using a rotating vane driven by an electric motor. The casing in which the **impeller** rotates also has an effect on the air movement characteristics. There are two basic configurations off an available each with different operating characteristics.

Axial flow fans have the impeller connected directly to the drive shaft of the motor (Fig.6.4). This means that the airflow passes over the motor and is parallel to the axis of the fan.

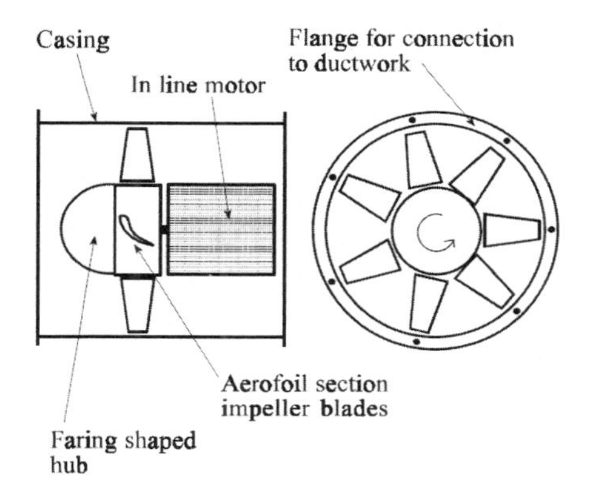

Fig.6.4 Axial flow fan.

The impeller blades can be one of two types. The first is a simple propeller where the blades have a uniform cross section throughout their length but are twisted so that when the impeller rotates the air that it comes into contact with is pushed from the leading to trailing edge of the blade. The momentum built up carries the air out

impeller
叶轮

of the fan. The volume of air moved depends on the speed of rotation of the fan and the number of blades. **Rotational** speed is however kept below 30 m/s blade tip velocity as the noise generated by the fan becomes unacceptable. Propeller fans do not generate a large pressure difference and so they cannot move air through ducting longer than approximately 45~55 cm. However they are effective at moving air through free openings such as window extract units or wall extract units incorporating short lengths of ducting.

The second blade arrangement is more complex having an aerofoil cross section and twisting from one end to the other. In the same manner as an aircraft wing, the aerofoil bladed impeller generates increased air movement over its upper surface. This creates a greater pressure difference and so the fan can move air along a system of ducting. The efficiency with which the aerofoil bladed fan converts electricity into air movement is higher than the propeller fan but can be further increased using stationary radial guide vanes across the inlet or exit of the fan. This reduces **swirl** and so gives a more even flow of air.

Centrifugal fans have a completely different impeller and casing arrangement to axial fans. The impeller blades rather than being perpendicular to the axis of rotation are parallel to it and are arranged into a drum like configuration (Fig.6.5). Air is drawn into the fan parallel to the axis of the

rotational
旋转速度，角速度

swirl
涡旋，湍流

impeller. The rotation of the impeller causes the air to leave the fan at right angles to the direction of entry. The air is driven by centrifugal force and collected by the volute casing. Changing the arrangement of blades within the impeller changes the characteristics of the fan. For slow, low pressure applications required in noise sensitive environments, the leading edge of the blades faces backwards, away from the direction of rotation. For maximum pressure generation required for moving air through very long lengths of ducting, the leading edge is made to face forwards.

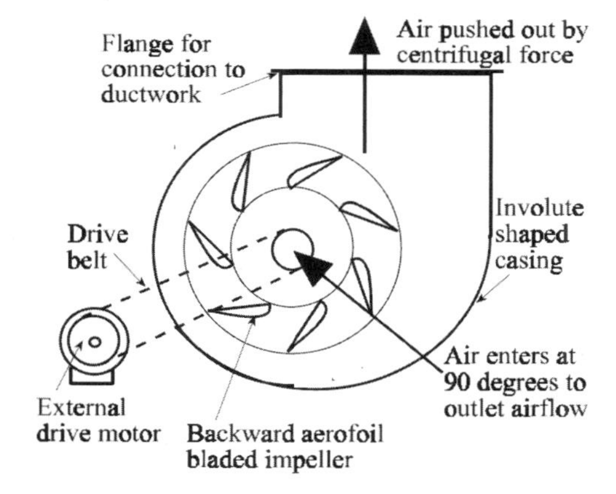

Fig. 6.5 Centrifugal fan.

6.5 Heat Recovery
热回收

Mechanical ventilation gives good control of

热回收即回收建筑物内外的余热（冷）或废热（冷），并把回收的热（冷）量作为供热（冷）或其他加热设备的热源而加以利用。

ventilation rates and hence air quality. However, the air removed from the building carries with it the energy used to warm it up to room temperature. It therefore represents a source of heat loss from the building. Mechanical ventilation has an advantage over natural ventilation in that systems can be put in place to recover most of the heat normally lost along with the extract air. The systems are known as air to air heat recovery units. Each method of heat recovery involves transferring energy from the exhaust airstream to the supply airstream. There are a number of ways of achieving this as described in the following sections.

Air to air plate heat exchangers are used for both domestic and commercial heat recovery. The remaining systems, because of their cost and complexity, are restricted to commercial use.

Plate heat exchangers are used where the exhaust and supply airstreams are arranged to flow alongside each other (Fig.6.6). They are composed of a cubical sandwich of thin metal or plastic plates. These plates allow the exhaust and supply airstreams to pass each other but remain separated. Heat passes from the hotter to the cooler airstream by conduction through the thin plates (Fig.6.7).

Plate heat exchangers have a number of advantageous features:

(1) They have no moving parts which would require maintenance.

(2) They keep the airstreams separate so no cross contamination can occur.

(3) No energy is required for their operation although fan power may need to be increased to overcome air friction through the unit.

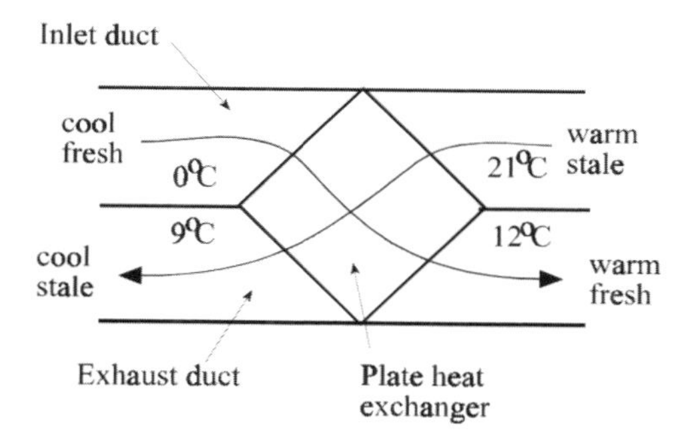

Fig.6.6 Location of plate heat exchanger in the ducting.

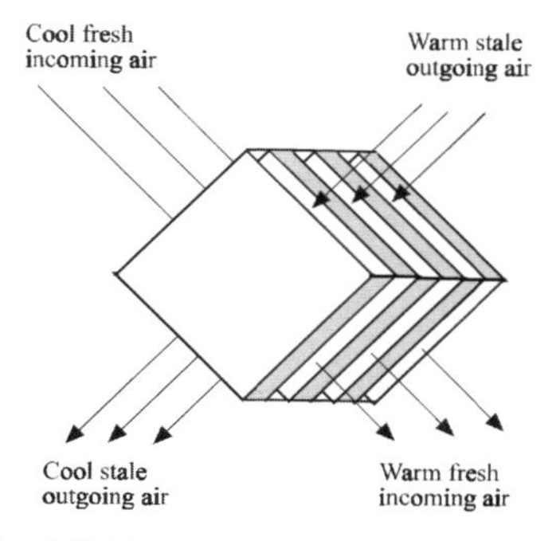

Fig. 6.7 Air to air plate heat exchanger.

Thermal wheels are composed of a circular matrix of tubes through which air can flow

(Fig.6.8). The wheel is positioned across the inlet and exhaust ducting so that inlet air passes through the upper half of the wheel and exhaust air passes through the lower half. As the exhaust air passes through the wheel the matrix heats up. The wheel rotates and slowly brings this heated section into the path of the incoming airstream. The incoming airstream is warmed as it passes through the thermal wheel matrix.

The thermal wheel requires an electric motor to drive it and so the energy consumption of this needs to be considered in assessing the heat recovery efficiency. Thermal wheels should not be used in areas where cross contamination of airflows would be a problem such as hospital operating theaters. This is because it is not possible to fully seal between supply and exhaust airflows.

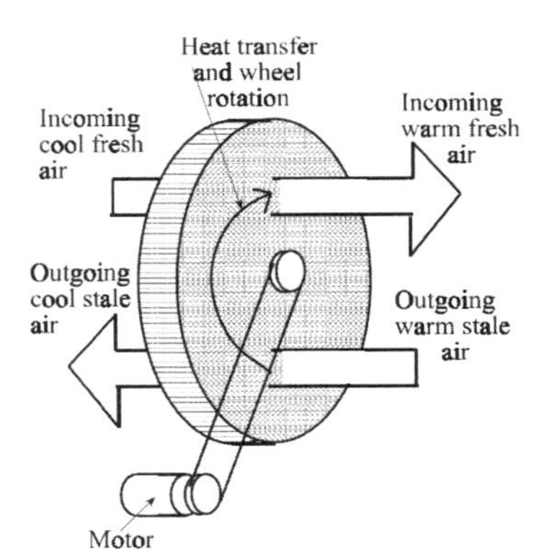

Fig. 6.8 Thermal wheel.

Run around coils are heat recovery devices which can be used when supply and exhaust airflows are not run close together. A finned coil is situated in the path of the exhaust air (Fig. 6.9). Air passing through the coil heats up a water and antifreeze mixture circulating in the pipe work. A pump circulates this heated solution to a similar coil installed in the supply duct. The supply air will become heated by passing through this coil. Topping up and maintenance of the system will need to be considered in economic analyses.

Heat pumps are similar to run around coils in that they can exchange heat between ducts which are separated by distance. However the pipe work is filled with a refrigerant which evaporates in the coil situated in the exhaust duct. As the refrigerant evaporates it absorbs heat. The refrigerant vapor then flows to the coil in the supply duct where it condenses and in doing so releases the heat it has absorbed. The pipe work is fitted with a compressor and pressure reducing valve which enable the evaporation and condensation of the refrigerant to take place.

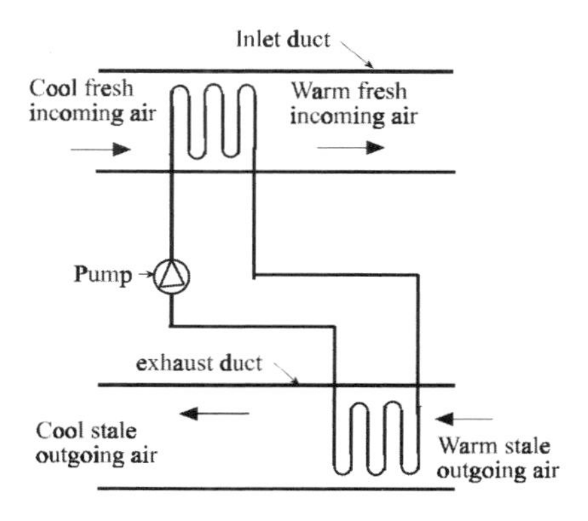

Fig. 6.9 Run around coil.

本章重点内容介绍

随着建筑的重要性日益上升，通风的地位变得举足轻重，所有居住空间都需要通风，以保持良好的空气质量和舒适的温度。当室内的气体成分所占比例不同时（比如二氧化碳），会对人体产生各种各样的影响（负面）。温度也是同理，建筑中的热量来自人体、电动设备、太阳辐射以及某些会直接散发热量的设备（如炉灶）。通过精确地设计、维护和运行，通风系统就可以创造出一个精确控制的舒适环境。不过也因此衍生出一些问题，比如设备的造价成本、管网的清洁以及设备运行噪声等等，因此，在通风领域充满了各种工程问题。

通风被分为两大类：自然通风和机械通风。其中，自然通风利用室内外空气的密度差以及空气的流动，在这一过程中存在不稳定性，因为空气的渗透是随机的，但是自然通风本身因为不消耗机械能，所以是几乎没有成本存在的，所以在对空气要求并不高的建筑物内，自然通风是最合理且经济性最高的“换气”方式。如果建筑物对空气品质的要求较高且精细（比如工艺生产车间），自然通风无法使空气达到产品对生产环境的需求，那么这个时候需要机械通风的介入，而机械通风比起自然通风要复杂许多，这其中要考虑不同层数（高度）以及面积和其他设备、人员对空气进行污染的情况。自然通风和机械通风各有优点，因此在设计建筑通风的时候，必须综合考虑这两类通风方式所涉及的影响因素。

Chapter 7
Air Conditioning
空气调节

7.1 Introduction
简　介

Air-conditioning (cooling) is refrigeration applied to the space temperature of a building to cool it during the hot summer months. The air-conditioning system removes the heat that leaks into the structure from the out-side and deposits it outside the structure where it came from. Some people living in the warmer sections of the country may never have air-conditioning, but at times, they are probably uncomfortable. When the nights are warm and the **humidity** is high, it is hard to be comfortable enough to rest well.

空气调节是对某一房间或空间内的温度、湿度、洁净度和空气流动速度等进行调节与控制，并提供足够量的新鲜空气。

The basics of refrigeration and its components were discussed in Chapter 4. Some of the material is covered again here so that it will be readily available. As you study this chapter, you will find that many of the components used

humidity
湿度、潮湿

for refrigeration applied to air-conditioning are different from the components applied to commercial refrigeration.

7.2 Structural Heat Gain
结构热增益

Heat leaks into a structure by conduction, **infiltration**, and radiation (the sun's rays or solar load). The summer solar load on a structure is greater on the east and west sides because the sun shines for longer periods of time on these parts of the structure (Fig.7.1). If a building has an attic, the air space can be ventilated to help relieve the solar load on the ceiling, in Fig.7.2(a) and Fig.7.2(b). There are two kinds of attic ventilation, power and natural. Fig.7.2(a) shows a power **ventilator**. It is a thermostatically controlled fan that starts operating when the attic temperature reaches a predetermined temperature, typically 30 °C. This fan must have a high-limit control to shut the fan off in case of an attic fire. The power ventilator would just feed air to a fire. Fig.7.2(b) shows a natural-draft ventilation system that is used more extensively. It is called a ridge vent system. The ridge vent is cut along the top of the entire roof ridge. Another vent, called a soffit vent, typically runs along the bottom of the roof overhang. These two vents create air circulation along the entire underside of the roof surface. Both systems are said to keep the roof

infiltration
渗透

ventilator
通风设备

shingles cooler and to prolong roof life. The power vent consumes some power and requires service when broken. If the structure has no attic, it is at the mercy of the sun unless it is well insulated (Fig.7.3).

Fig. 7.1 The solar load on a home.

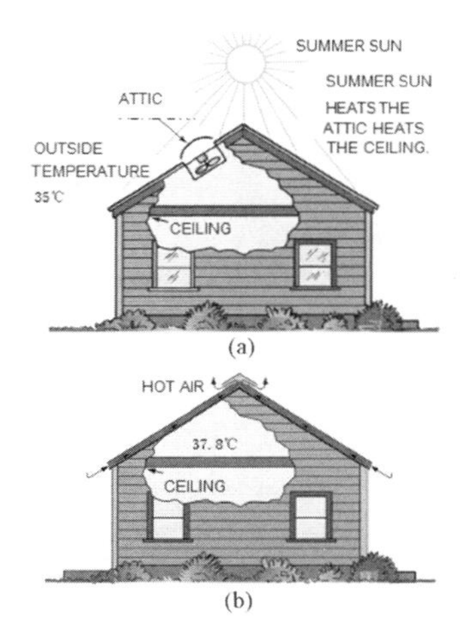

Fig. 7.2 A ventilated attic helps keep the solar heat from the ceiling of the house.

Conduction heat enters through walls, windows, and doors. The rate depends on the temperature difference between the inside and the outside of the house (Fig.7.4). Some of the warm air that gets into the structure infiltrates through the cracks around the windows and doors. Air also leaks in when the doors are opened to allow people to enter and leave the building. This infiltrated air has different characteristics in different parts of the country. Using the example in Fig. 7.4, the typical design condition in Phoenix is 40.6 ℃ dry-bulb and 21.7 ℃ wet-bulb. In Augusta, the air may be 32.2 ℃ dry-bulb and 22.8 ℃ wet-bulb. When the air leaks into a structure in Phoenix, it is cooled to the space temperature. This air contains a certain amount of humidity for each cubic foot that leaks in. In Augusta, there will normally be more humidity with the infiltration than in Phoenix.

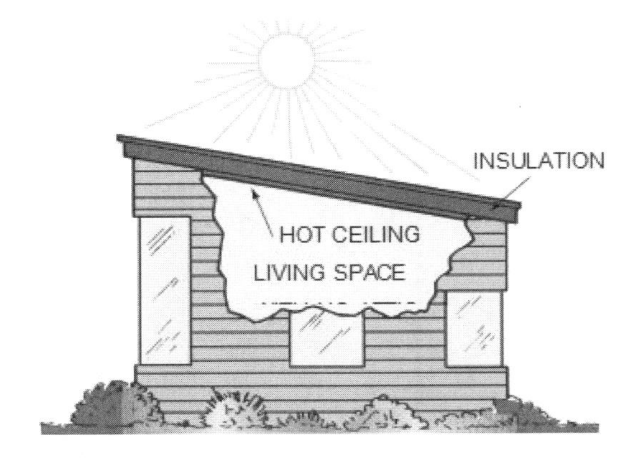

Fig.7.3 This house has no attic. The sun shines directly on the ceiling of the living space.

The humidity difference in two different locations must be accounted for in the choice of conditioning equipment. Equipment selected for parts of the country that are humid will have more capacity to remove moisture from the conditioned space. For this reason, used equipment cannot just be readily moved from one locality to another.

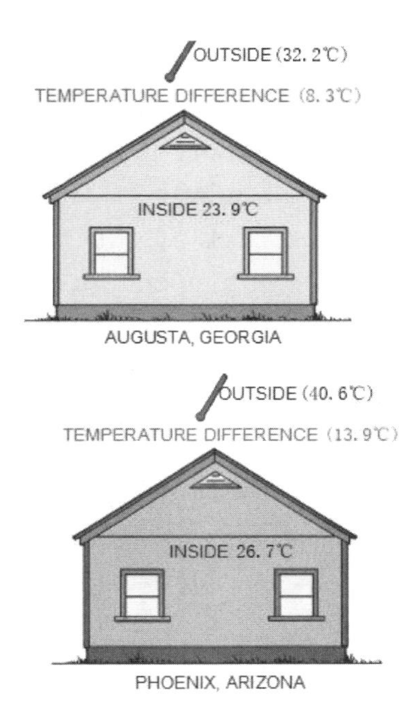

Fig. 7.4 The difference between the inside and outside temperatures of a home in Augusta, Georgia, and a home in Phoenix, Arizona.

7.3 Evaporative Cooling

蒸发冷却

Air has been conditioned in more than one way

蒸发冷却是利用液体蒸发过程从周围吸热，从而实现冷却功能。这种物理现象早为人类所利用，例如在房间内泼点水，可以使室温有所下降。

to achieve comfort. In the climates where the humidity is low, a device called an evaporative cooler (Fig.7.5), has been used for years. Water slowly running down the fiber mounted in a frame is the cooling media. Fresh air is drawn through the water-soaked fiber and cooled by evaporation to a point close to the wet-bulb temperature of the air. The air entering the structure is very humid but cooler than the dry-bulb temperature. For example, in Phoenix, Arizona, the design dry-bulb temperature in summer is 40.6 °C. At the same time the dry-bulb is 40.6 °C, the wet-bulb temperature may be 21.1 °C. An evaporative cooler may lower the air temperature entering the room to 26.7 °C dry-bulb, which is cool compared to 40.6 °C, even if the humidity is high.

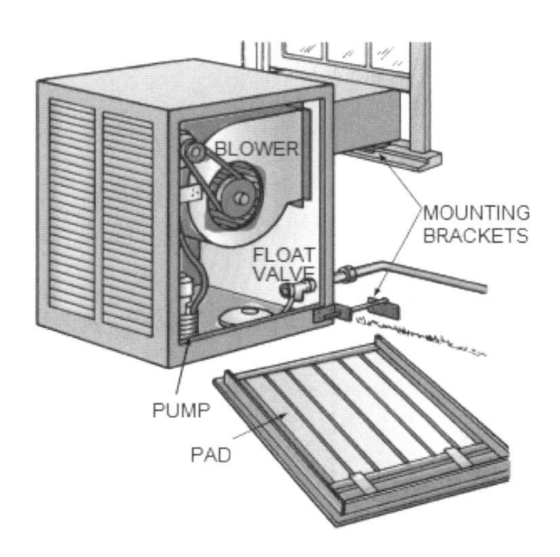

Fig. 7.5 An evaporative cooler.

Evaporative cooling units use 100% outdoor air. There must be an outlet through which the air that enters the structure can leave it. The conditioned air will have a tendency to move toward the outlet used to exhaust the air from the structure. Opening a window on the far side of the structure from where the air enters allows air to exit; otherwise, the structure will become pressurized by air that is being pushed in.

7.4 Refrigerated Cooling and Air Conditioning
制冷与空调

Refrigerated air-conditioning is similar to commercial refrigeration because the same components are used to cool the air: (1) the evaporator, (2) the compressor, (3) the condenser, and (4) the metering device. The four components of refrigerated air-conditioning are assembled into two basic types of equipment for air-conditioning purposes: package equipment and **split-system** equipment. These components are assembled in several ways to accomplish the same goal — refrigerated air to cool space.

1. Package Air-Conditioning

With package equipment all of the components are built into one cabinet. It is also called self-contained equipment (Fig.7.6). Air is ducted to

split-system
分裂系统

and from the equipment. Package equipment may be located beside the structure or on top of it. In some instances, the heating equipment is built into the same cabinet.

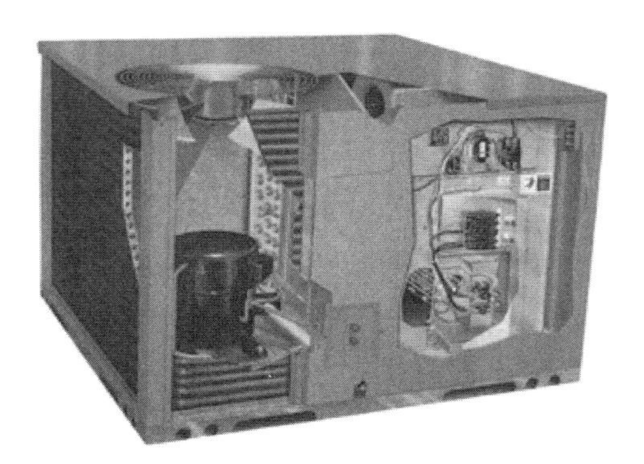

Fig. 7.6 Package air conditioner.

2. Split-System Air-Conditioning

In split-system air-conditioning, the condenser is located outside, remote from the evaporator, and uses interconnecting refrigerant lines. The evaporator may be located in the attic, a crawl space, or a closet for upflow or downflow applications. The blower that moves air across the evaporator may be included in the heating equipment, or a separate blower may be used for the air-conditioning system (Fig.7.7).

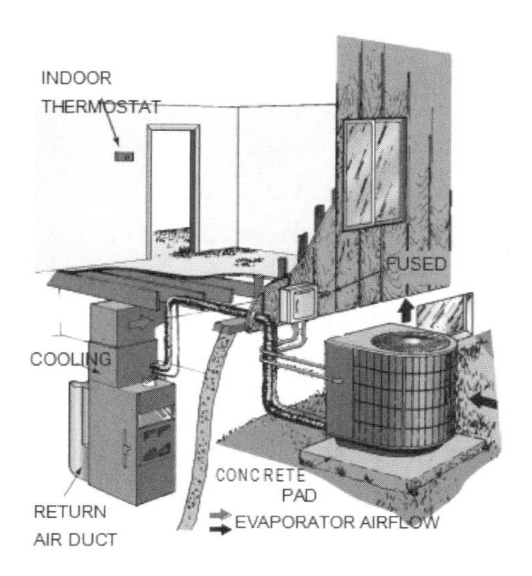

Fig. 7.7 A split air-conditioning system.

7.5 The Evaporator
蒸发器

The evaporator is the component that absorbs heat into the refrigeration system. It is a refrigeration **coil** made of aluminum or copper with aluminum fins attached to the coil of either type to give more surface area for better heat exchange. There are several designs for airflow through the evaporator coil and for draining the condensate water from the coil, depending on the installation. The different designs are known as the A coil, the slant coil, and the H coil.

coil
线圈，盘管

7.6 The Function of the Evaporator 蒸发器的作用

The evaporator is a heat exchanger that takes the heat from the room air and transfers it to the refrigerant. Two kinds of heat must be transferred: sensible heat and latent heat. Sensible heat lowers the air temperature, and latent heat changes the water vapor in the air to condensate. The condensate collects on the coil and runs through the drain pan to a trap (to stop air from pulling into the drain), and then it is normally piped to a drain.

Typically, room air may be 23.8 ℃ **dry-bulb** and have a humidity of 50%, which is 16.9 ℃ wet-bulb. The coil generally operates at a refrigerant temperature of 4.4 ℃ to remove the required amount of sensible heat (to lower the air temperature) and latent heat (to remove the correct amount of moisture). The air leaving the coil is approximately 12.8 ℃ dry-bulb with a humidity of about 95%, which is 12.2 ℃ wet-bulb. Notice the high humidity leaving the coil. This is because the air has been cooled. When it mixes with room air, it is heated and expands where it can absorb moisture from the room air. The result is dehumidification.

These conditions are average for a climate with high humidity. If the humidity is very high, such as in coastal locations, the coil temperature

dry-bulb
干球的

may be a little lower to remove more humidity. If the system is in a locality where the humidity is very low, such as in desert areas, the coil temperature may be a little higher than 4.4 ℃. The coil temperature is controlled with the airflow across the evaporator coil. More airflow will cause higher coil temperatures, and less airflow will cause lower coil temperatures. The same equipment can be sold in all parts of the country because the airflow can be varied to accomplish the proper evaporator temperature for the proper humidity. Fin **spacing** on the evaporator may also affect the leaving air's humidity levels. Closer fin spacing will take more humidity out of the air.

7.7 Indoor Air Quality 室内空气品质

Inadequate ventilation is a significant problem associated with poor indoor air quality (IAQ). Ventilation is the process of bringing fresh clean air into the building and exhausting stale indoor air out of the building. This is often done by a heating, ventilation and air-conditioning system called an HVAC system. The HVAC system controls the temperature of a building as well as the humidity levels. In addition to controlling temperature and humidity, HVAC systems can also control odors and dilute pollutants inside the building.

室内空气品质反映了居住者的满意程度。民用建筑室内空气品质的优劣，传统上用几种污染物的浓度作为控制指标。

spacing
间隔

Some studies estimate that approximately 60% of all indoor air problems are due to ventilation. Of this 60%, 30% are a result of air contamination, 20% from inside the building and 10% from outside. If there is not good ventilation, pollutants can build up inside the building and result in indoor air complaints. A NIOSH (National Institute of Occupational Safety and Health) study blamed about half the cases of indoor air pollution on poor ventilation. According to a report from the US Accounting Office, an estimated 20 percent of all schools in the United States have indoor air problems and 25% have unsatisfactory ventilation. In addition, 36% of all school officials rated their facility's heating and ventilation systems as a "less than adequate building **feature**".

Having good ventilation systems can decrease the number of indoor air complaints and health problems caused by poor air quality. A study by the Walter Reed Institute of Medical Research found that absentee rates in poorly ventilated buildings are 50% greater than in well-ventilated buildings. Several studies have shown that the number of Sick Building Syndrome (SBS) is 40%~50% lower among office workers 6 months after moving into buildings with good ventilation systems. In a study in Quebec City, researchers showed that this large decrease in the number of SBS complaints was maintained 3 years later. In this study, one of the reasons for the fewer SBS

feature
特点、特征

problems was that in the new building the copying machines were in separate rooms away from workers. Another reason was the smoking area had its own exhaust system so that contaminated air was not recirculated throughout the building. Nevertheless, poor ventilation can also affect the quality of work that one does. In a recent Danish study, researchers found that typists in offices with clean air had 6% more output than typists in other offices.

There are many different parts to an HVAC system. These parts include furnaces or boilers, chillers, cooling towers, air handling units, exhaust fans, ductwork, filters, steam (or heating water) piping, dampers, louvers, cooling coils, drip pans, registers, and grills. All of these parts need to be kept clean, free of debris and mold, and in good working condition in order to provide good quality air to the building. In addition, it is important that the location of the air intake units be considered. It is possible for rooftop units to pull pigeon and other bird droppings into the HVAC system as part of the outside "fresh" air brought into the building. This increases the chance of people inside the building developing lung infections such as histoplasmosis or psittacosis. Intake units located near building exhausts, parking lots, loading docks, or busy highways increase the chance of bringing into the building vehicular exhaust, carbon monoxide, VOCs, and particulates. Intakes located near or downwind of cooling towers increase the risk of legionella **bacteria** entering the air supplies of buildings.

7.8 Ventilation Rates for Offices and Schools 办公室和学校的通风率

Ventilation rates are measured in cubic meters per hour and guidelines are written in m^3/h per person. The guidelines that are often used for ventilation rates are provided by an organization called the American Society of Heating, Refrigerating and Air-Conditioning Engineers (ASHRAE). ASHRAE recommendations are voluntary standards that only become enforceable when a state or locality adopts their standards in their building codes. In the United States, there are no federal laws requiring a minimum amount of fresh airflow into buildings.

ASHRAE guidelines have changed over the years. One of the fist recommendations to improve indoor air quality was in 1836 when Tregold calculated the minimum amount of outdoor air needed for miners to be 6.78 m^3/h. In 1895 ASHRAE adopted a recommendation of 50 m^3/h to reduce the spread of disease, especially tuberculosis. In the early and mid-1900s, building ventilation standards called for about 25 m^3/h of outside air for each building occupant. The

bacteria
细菌

purpose at that time was to mainly reduce body odors in the building.

Early ventilation recommendations for schools were first proposed in the early 1900s in Massachusetts and New York. These states recommended 50 m^3/h per student. States began realizing that by bringing in fresh outdoor air, they were able to reduce the spread of respiratory infections.

Because of the 1973 oil embargo, however, the amount of outdoor air for ventilation was reduced to 8.5 m^3/h per occupant. This was considered necessary to save energy as long as there were no major sources of pollutants. Another action that was taken to conserve energy around that time was that buildings were "tightened up" to prevent leaks and new buildings were built with windows that did not open. This tightening of the building along with reduced ventilation rates began causing discomfort and health problem in building occupants. The pollutants inside had no way to get out and began to build up. This became as Sick Building Syndrome (SBS).

ASHRAE currently recommends 34 m^3/h of outdoor air per person for office spaces and conference rooms, and at least 25 m^3/h of outdoor air for other nonresidential environments such as school classrooms. However, this may not be enough. In a study of 399 workers in 14 chanically ventilated office buildings in Helsinki, Finland, the researchers found that outdoor air ventilation rates below 54~90 m^3/h increases the risk of symptoms of SBS. In France, as compared to the United States, buildings are required to have 3 times more ventilation, 2 times more soundproofing, while using half as much energy.

However, it is not enough to just increase the amount of outdoor air. In a study of 1,838 buildings, the ventilation rates were increased from 34 m^3/h to 85 m^3/h for 2 weeks without the occupants knowing about the increase. However, this increase in ventilation rates did not change the workers' perceptions of the air in their offices or decrease the number of SBS complaints received. Another well-controlled study has shown that increasing the supply of outdoor air does not affect workers' perceptions of their office environment or their reporting of symptoms considered typical of building-related illness. From these studies, it would appear that increasing the ventilation rates does not make any difference. However, there is a variety of factors that need to be considered such as the layout of the space. For example, offices that use partitions to divide the space into individual cubicles are also creating barriers to adequate air reaching everybody. The activities going on inside the building are also important since indoor air pollutants are increased by activities such as photocopying, smoking, or handling paper.

7.9 Problems with Ventilation Systems 通风系统的问题

Often problems with ventilation systems occur because the building is being used for a different purpose or has more people in it than were originally thought of when it was designed. First, look at these factors to see if they might explain some of the problems that may be happening:

(1) How was the building originally intended to function? For example, if the building was originally built as a factory with an open design, there may be ventilation problems if that open space is broken up walls and partitions. The activities going on in the building may also be different than originally conceived.

(2) Is the building functioning as designed?

(3) What changes in building layout and use have occurred since the original design and construction? Find out if the HVAC system has been reevaluated to make sure it reflects the current use of the building.

When there are changes, there may be indoor air quality problems. For example, there was one building built in the early 1960s in Bonn, Germany that consisted of three naturally ventilated sections. Until the mid-1980s, the building was used by a pharmaceutical company. Some parts of this building were used for production and storage of pharmaceutical products. When a new company took over the building, they converted this production and storage space into offices. Later there were complaints of indoor air problems. Upon investigation, researchers found that the people who were complaining about indoor air problems were in the same areas that had at one time been the production and storage areas. These areas had been converted to office space without regard to the prior use of these parts of the building.

When one study looked at HVAC problems in several buildings, they found the following problems:

(1) 70% had inadequate amounts of outside air.

(2) 50%~70% had inadequate air distribution to the parts of the buildings where people worked.

(3) 60% had inadequate filtration to remove outdoor pollutants.

(4) 60% had standing water in the system which had the potential for microbial growth.

(5) 40% with visible mold and fungus growing on the insulation.

(6) 20% with malfunctioning humidifies.

(7) 75% had inadequate maintenance programs.

Mechanical ventilation systems in large buildings are designed and operated not only to heat and cool the air, but also to draw in and circulate outdoor air. If they are poorly designed, operated, or maintained, however, ventilation systems can create indoor air problems. Ventilation problems are the biggest cause of indoor air quality problems:75% do not get enough outside air; 65% have problems distributing the air to all occupants (partitions and other obstacles that may block supply and return vents); 60% have improper filtration; 75% have inadequate maintenance of the air handling system; and 45% have microbial contamination. Some of these problems are due to poor initial design, outdoor vents or dampers closed to "save energy", inadequate maintenance of dampers and control linkages, and obstructed outdoor screens on the air intakes. There may also be fans that have been inoperative for a number of years or filters that have not been changed on a regular basis. Some of the other problems in the ventilation equipment itself include:

(1) Condensate in pans and ductwork. This can lead to microbial growth such as bacteria, slime, algae, and mold.

(2) **Fiber glass** being shed by an interior duct lining. Fiber glass was used as a liner in buildings

fiber glass
玻璃纤维

in the late 1970s and the early 1980s in order to reduce noise from the HVAC fans and for thermal insulation. After several years, these linings begin releasing fibers. For example, there was an office in Richmond, Virginia in which workers thought they were suffering from insect bites for several months. When somebody was brought in to check out the problem, they found that some filters were missing and loose fiber glass was being blown through the vents onto workers. As soon as the filters were installed, their skin cleared up.

(3) Fiber glass that is dirty and becomes wet can be a source of mold growth. Mold spores from this growth can become airborne and distributed throughout the occupied spaces of buildings.

(4) Broken dampers can prevent outdoor air from being brought into the building.

(5) Pneumatic controls that do not work properly can determine whether a sufficient amount of outdoor air is being brought into the building.

(6) Poor quality filters can reduce the efficiency of the mechanical system in distributing air. They can also allow dirt, particles, pollen and other debris to enter the building. Filters that become clogged with dirt can become sources of mold and bacterial growth.

本章重点内容介绍

空气调节（制冷）是在炎热的夏季对建筑物的空间温度进行制冷以使其降温。空调系统将从外部进入到建筑物中的热量移除，并将其热量沉积在建筑外部。

热通过传导、渗透和辐射(太阳射线或太阳负载)泄漏到建筑物中。太阳辐射引起向建筑物内传递的热量负荷更大，因为太阳在建筑物的这些部分上照射更长的时间。如果建筑物有阁楼，空气空间可以通风，以帮助减轻天花板上的太阳能负荷。阁楼通风分为机械通风和自然通风两种。动力通风机是一种恒温控制风扇，在剧院温度达到预定温度时开始工作，通常为 30 英寸。

制冷空调类似商用制冷，因为相同的部件被用来冷却空气：(1) 蒸发器，(2) 压缩机，(3) 冷凝器，(4) 计量装置。制冷空调的四个部件被组装成两种基本类型的空调设备：包装设备和分体式空调设备。这些部件以多种方式组装，以实现同样的目标——制冷空气以冷却空间。

有了包装设备，所有的组件都建在一个柜子里。它也被称为自给设备。空气通过管道进出设备。包装设备可能位于结构旁边或顶部，在某些情况下，加热设备被安装在同样的橱柜里。

在分体式空调系统中，冷凝器位于外面，远离蒸发器，使用相互连接的制冷剂管路。蒸发器可能位于阁楼爬行空间，或用于向上或向下流动应用的壁橱。使空气流过蒸发器的鼓风机可能包括在加热设备中，或者一个单独的鼓风机可用于空调系统。

蒸发器是将热量吸收到热力系统中的部件。它是一种由铝或铜制成的制冷盘管，铝翅片附着在另一类型的盘管上，为更好地换热提供了更多的表面积。蒸发器盘管中的气流和从盘管中排出冷凝水的设计有几种形式，具体形式取决于安装方式。不同的设计有 A 线圈、斜线圈和 H 线圈。

蒸发器是一种热交换器，它把房间空气中的热量输送给制冷剂。必须传递两种热量：显热和潜热。显热降低空气温度，潜热使空气中的水蒸气凝结。这些凝析物聚集在线圈上，穿过排水管，形成一个陷阱（阻止空气进入排水沟），然后通常用管道输送到排水沟。

通风不足是与室内空气质量差有关的一个重要问题。通风是将新鲜洁净的空气带入建筑物，排出室内空气的过程。这通常是由一个称为暖通空调系统的供暖、通风和空调系统完成的。暖通空调系统以湿度水平来控制建筑物的温度。除了控制温度和湿度外，暖通空调系统还可以控制建筑物内的污染物并稀释污染物。

Chapter 8
Air Source Heat Pump
空气源热泵

8.1 Introduction
简　介

Heat pumps, like refrigerators, are refrigeration machines. Refrigeration **involves** the removal of heat from a place where it is not wanted and depositing it in a place where it makes little or no difference. The heat can actually be deposited as heat reclaim in a place where it is wanted. This is the difference between a heat pump and a cooling-only air-conditioning system. The air conditioner can pump heat only one way. The heat pump is a refrigeration system that can pump heat two ways. Since it has the ability to pump heat into as well as out of a structure, the heat pump system can provide both heating and cooling. All compression-cycle refrigeration systems are heat pumps in that they pump heat-laden **vapor**. The **evaporator** of a heat pump absorbs heat into the refrigeration system, and the **condenser** rejects the heat. The compressor pumps the heat-laden vapor. The metering device controls the

空气源热泵是一种利用高位热能使热量从低位热源流向高位热源的节能装置。它是热泵的一种形式。可以把不能直接利用的低位热能转换为可以利用的高位热能，从而达到节约部分高位热能的目的。

involve
涉及

vapor
水蒸气

evaporator
蒸发器

condenser
冷凝器

refrigerant flow. These four components — the evaporator, the condenser, the compressor, and the metering device — are essential to compression-cycle refrigeration equipment. The four-way valve in a heat pump system is used to switch the unit between the heating and cooling modes of operation. By changing the mode of operation of the system, the functions of the indoor and outdoor coils change as well. In the cooling mode, the indoor coil acts as the evaporator and the outdoor coil functions as the condenser. In the heating mode, the indoor coil functions as the condenser and the outdoor coil operates as the evaporator.

8.2 Heat Sources for Winter 冬季热源

The cooling system in a typical residence absorbs heat into the refrigeration system through the evaporator and rejects this heat to the outside of the house through the condenser. The house might be 24 °C inside while the outside temperature might be 35 °C or higher. The air conditioner pumps heat from a low temperature inside the house to a higher temperature outside the house; that is, it pumps heat up the temperature scale. A **freezer** in a supermarket takes the heat out of ice cream at –17.8 °C to cool it to –23.3 °C so that it will be frozen hard. The heat removed can be felt at the condenser as hot

freezer
冷冻室

air. This example shows that there is usable heat in a substance even at –17.8 °C (Fig.8.1). There is heat in any substance until it is cooled down to –273.15 °C (0 K).

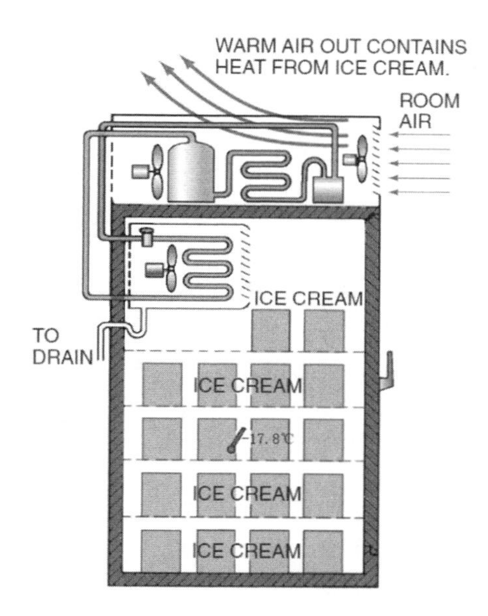

Fig 8.1 This low-temperature refrigerated box is removing heat from ice cream at -17.8 °C. Part of the heat coming out of the back of the box is coming from the ice cream.

If heat can be removed from –17.8 °C ice cream, it can be removed from –17.8 °C outside air. The typical heat pump does just that. It removes heat from the outside air in the winter and deposits it in the conditioned space to heat the house. (Actually, about 85% of the usable heat is still in the air at –17.8 °C.) Hence, it is called an air-to-air heat pump (Fig.8.2). In summer, the heat pump acts like a conventional air conditioner and removes heat from the house and deposits it outside. From the outside an air-to-air heat pump looks like a central cooling air conditioner (Fig.8.3). The phrase "air to air" indicates the source of heat while the system is operating in the heating mode and the medium that is ultimately being treated. The first "air" represents the heat source, and the second "air" represents the medium being heated. For example, an "air-to-water" heat pump uses "air" to heat "water". Water-source heat pumps use water as the heat source in the heating mode and can be "water to water" or "water to air". The water-to-air heat pump, therefore, uses water as the heat source to heat air.

The water source can be, for example, lakes or wells.

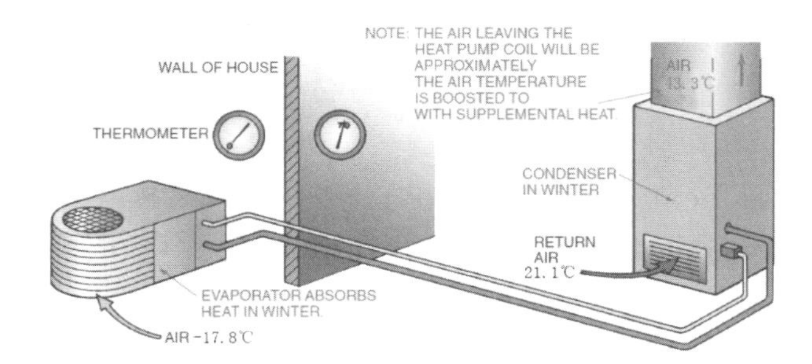

Fig. 8.2 An air-to-air heat pump removing heat from -17.8 °C air and depositing it in a structure for winter heat.

Fig. 8.3 The outdoor unit of an air-to-air heat pump. It absorbs heat from the outside air for use inside the structure.

8.3 The Four-Way Reversing Valve

四通换向阀

The refrigeration principles that a heat pump uses are the same as those stated previously. However, a new component is added to allow the refrigeration equipment to pump heat in either direction. The air-to-air heat pump in Fig. 8.4 shows the heat pump moving heat from inside the conditioned space in summer to the outside. Then in winter the heat is moved from the outside to the inside (Fig.8.5). This change of direction is accomplished with a special component called a four-way reversing valve (Fig.8.6). This valve can best be described in the following way. The heat absorbed into the refrigeration system is pumped through the system with the compressor. The heat is contained and concentrated in the discharge

四通换向阀：通过改变制冷剂的流动通道，改变制冷剂流向，具有转换冬夏两季空调系统室内机和室外机的功用。夏季，制冷剂液体在室内机（此时为蒸发器）内蒸发吸热成为气体，在室外机（此时为冷凝器）中放热，用于室内供冷；冬季，制冷剂液体在室外机（此时为蒸发器）中蒸发吸收外界热量，在室内机（此时为冷凝器）中放热，用于室内供热。

gas. The four-way valve diverts the discharge gas and the heat in the proper direction to either heat or cool the conditioned space. This valve is controlled by the space temperature thermostat, which positions it to either HEAT or COOL.

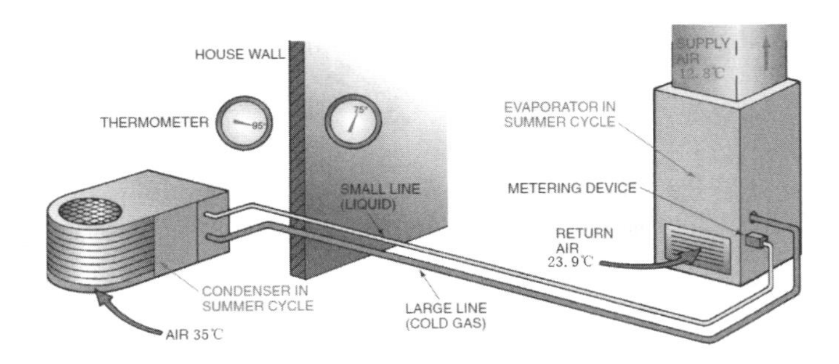

Fig. 8.4 An air-to-air heat pump moving heat from the inside of a structure to the outside.

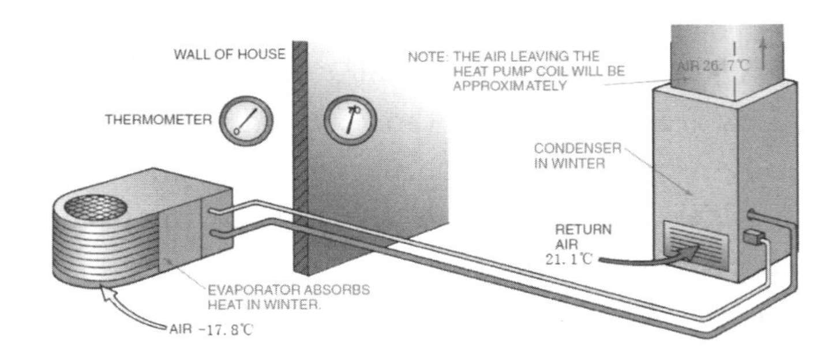

Fig. 8.5 In the winter, the heat pump moves heat into the structure.

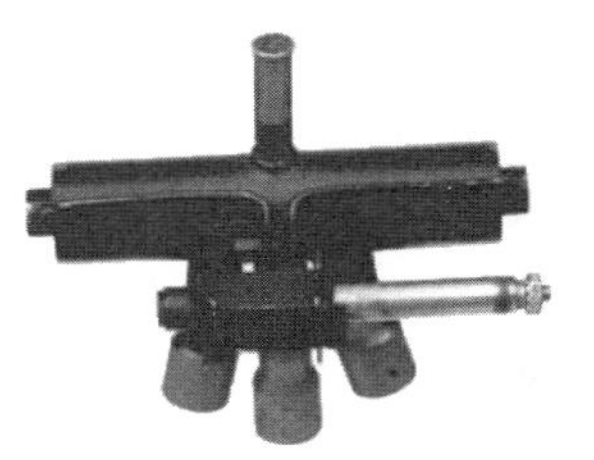

Fig. 8.6 A four-way valve.

The four-way reversing valve is a four-port valve that has a slide mechanism in it (Fig.8.7). The position of the slide is **determined** by a solenoid that is energized in one mode of operation and not the other. The position of the slide also determines the mode of system operation. Most heat pump systems are designed to operate in the heating mode when the solenoid is not energized (Fig.8.7[a]). This is referred to as failing in the heating mode. For the remainder of this text it will be assumed that the heat pump system will operate this way. This may not always be the situation, so be sure to check each individual system being worked on. Fig.8.7(b) shows the position of the slide when the solenoid is energized and the system is operating in the cooling mode.

Typical reversing valves have one port isolated on one side and the remaining three ports on the other. The isolated port is where the hot gas from the compressor enters the valve. Fig.8.7 shows the piping connections on the valve. Depending on the position of the slide, the hot gas will be directed to either the outdoor coil or the indoor coil. In either case, the coil that accepts the hot gas from the compressor is functioning as the condenser coil. Note that in (Fig.8.7[a]) the hot gas enters the valve at the top and is directed to the right-hand port, which is connected to the indoor coil. Since the indoor coil is receiving the hot gas from the compressor, the indoor coil is the condenser and the system is operating in the heating mode.

determine
决定

Compare this with (Fig.8.7[b]), where the hot gas from the compressor is being directed to the outdoor coil. Under these conditions, the system is operating in the cooling mode.

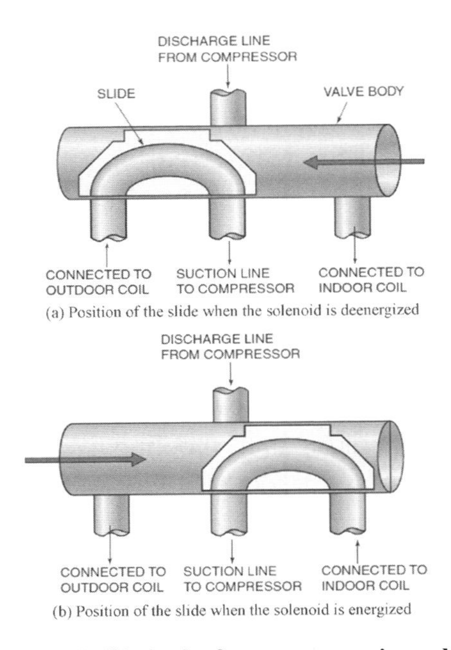

Fig. 8.7 Internal slide in the four-way reversing valve. The solenoid determines the position of the slide. Most systems operate in the heating mode when the solenoid is deenergized.

Smaller heat pump systems utilize smaller, direct-acting, four-way reversing valves. "Direct acting" implies that the solenoid itself provides the force needed to change the position of the slide in the valve. On larger systems, additional help is needed to move the slide. Valves in these systems, referred to as pilot-operated reversing valves (Fig.8.8), use the pressures in the heat pump system to help move the slide. The pilot-operated four-way valve is actually two valves built into one housing. The control wires control a small solenoid valve that has very small piping that runs along the outside of the valve, as can be seen in the bottom right-hand corner of Fig. 8.8. This valve is called a pilot-operated valve because a small valve controls the flow in a large valve.

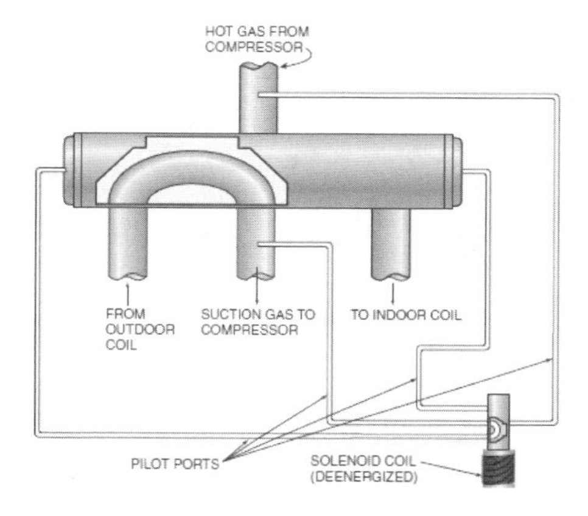

Fig. 8.8 A pilot-operated four-way reversing valve. Notice the pilot valve in the bottom right-hand corner of the figure.

The valve movement is controlled by pressure differences on a piston within the large portion of the valve. This piston slides back and forth on a platform inside the large valve. When the solenoid is deenergized, high pressure from the compressor discharge pushes through the tubes on the pilot valve as shown in Fig. 8.9. This exerts a pressure on the right side of the slide, pushing it to the left. The refrigerant on the left side of the slide is pushed back through the pilot valve and into the compressor's suction line. In this position, the hot gas from the compressor is directed to the indoor coil, putting the system into the heating mode. When the solenoid valve is energized, the pressure is then applied to the other end of the piston and the valve slides to the right for the cooling mode (Fig.8.10). This is why when starting up a heat pump, you sometimes hear the valve change position as the pressure difference is developed in the system.

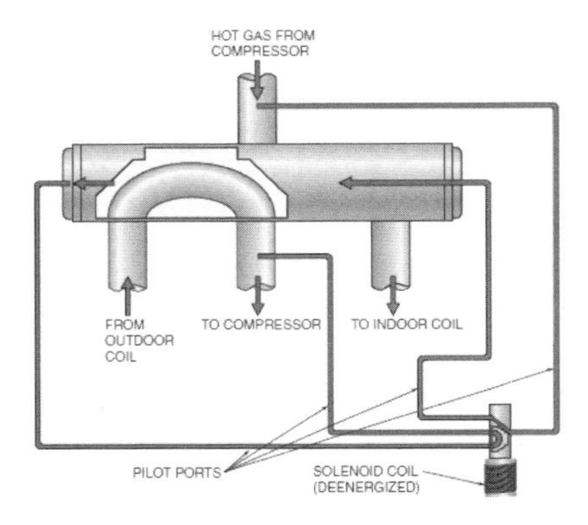

Fig. 8.9 In the heating mode, the high-pressure hot gas from the compressor is directed to push the main valve's slide to the left.

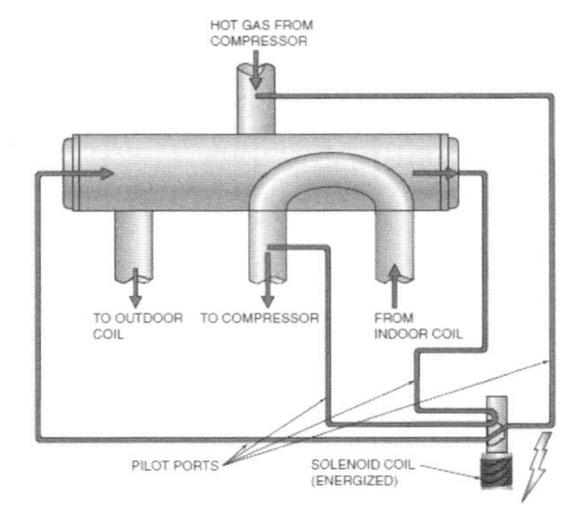

Fig. 8.10 In the cooling mode, the high-pressure hot gas from the compressor is directed to push the main valve's slide to the right.

8.4 The Air to Air Heat Pump 空气－空气热泵

The air-to-air heat pump resembles the central air-conditioning system. Both have indoor and outdoor system components. When discussing typical air-conditioning systems, these components are often called the evaporator (indoor unit) and the condenser (outdoor unit), but this terminology does not work for a heat pump system.

The coil that serves the inside of the house in a heat pump system is called the indoor coil. The unit outside the house contains the outdoor coil. The reason for the change in terminology is that the indoor heat pump coil is a condenser in the heating mode and an evaporator in the cooling mode. The outdoor coil is a condenser in the cooling mode and an evaporator in the heating mode, all of which is determined by the direction the hot gas is flowing. In winter the hot gas is flowing toward the indoor unit and will give up heat to the conditioned space. The heat must come from the out-door unit, which is the evaporator. See Fig. 8.11 for an example of the direction of the gas flow. The system mode can be determined easily by gently touching the insulated gas line to the indoor unit.

SAFETY PRECAUTION: If it is hot (it can be more than 90 °C, so be careful), the unit is in the heating mode.

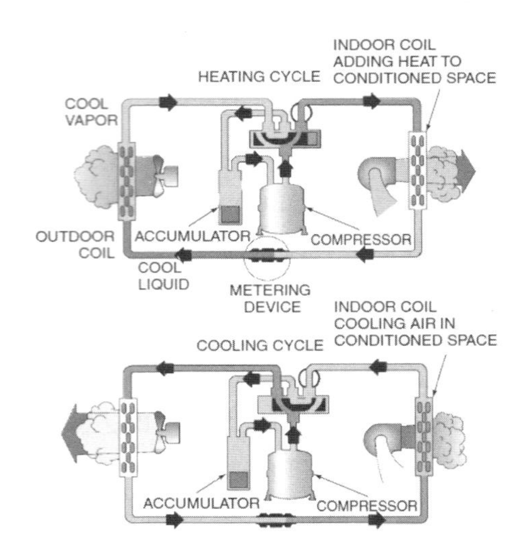

Fig. 8.11 The heat pump refrigeration cycle shows the direction of the refrigerant gas flow.

本章重点内容介绍

热泵是一种制冷系统，可以通过两种方式来输送热量。由于它能将热量泵入和泵出一个结构，热泵系统可以提供加热和冷却功能。所有的压缩循环制冷系统都是热泵，因为它们能泵送充满热量的蒸气。蒸发器、冷凝器、压缩机和计量装置对压缩循环制冷设备至关重要。热泵的蒸发器吸收热量进入制冷系统，冷凝器排出热量。压缩机泵送充满热量的蒸气。计量装置控制制冷剂流量。热泵系统中的四通阀用于在加热和冷却操作模式之间切换机组。通过改变系统的运行方式，室内和室外线圈的功能也随之改变。在冷却模式下，室内盘管作为蒸发器，室外盘管作为冷凝器。在加热模式下，室内盘管作为冷凝器，室外盘管作为蒸发器。

典型住宅中的冷却系统通过蒸发器将热量吸收到制冷系统中，并通过冷凝器将这些热量排出室外。冬天，它从室外空气中排出热量，并将其储存在有空调的空间中，为房子供暖。夏天，热泵的作用就像一个传统的空调，从房子里排出热量并将其沉积在外面。

热泵增加了一个新的部件，使制冷设备可以向任何方向泵送热量。热泵将夏季空调空间内的热量转移到室外。然后在冬季，热量从外部转移到内部，这种转向是用一种叫作四通换向阀的特殊部件来完成的。吸收到制冷系统中的热量由压缩机通过系统泵送。热量被包含并集中在排出的气体中。四通阀将排出的气体和热量转移到适当的方向，以加热或冷却经过调节的空间。该阀门由空间温度恒温器控制，恒温器将阀门设置为加热或冷却。

空气对空气的热泵类似于中央空调系统。两者都有室内和室外系统组件。在热泵系统中为室内服务的线圈称为室内线圈。屋外的装置包含室外线圈。室内热泵盘管在加热模式下是冷凝器，在冷却模式下是蒸发器。室外盘管是处于冷却模式的冷凝器还是处于加热模式的蒸发器，取决于热气的流动方向。在冬季，热气体流向室内机，并将热量释放到一定的空间。热量必须来自室外机，即蒸发器。轻触室内机的绝缘燃气管道，即可轻松确定系统模式。

Chapter 9
Geothermal Heat Pumps
地源热泵

地源热泵是陆地浅层能源通过输入少量的高品位能源（如电能等）实现由低品位热能向高品位热能转移的装置。根据地热能交换系统形式的不同，地源热泵系统分为地埋管地源热泵系统、地下水地源热泵系统和地表水地源热泵系统。通常地源热泵消耗 1 kW·h 的能量，用户可以得到 4 kW·h 以上的热量或冷量。

Geothermal heat pumps are refrigeration machines. They are very similar to air source heat pumps in that they can remove heat from one place and transfer it to another. However, geothermal heat pumps use the earth, or water in the earth, for their heat source and heat sink. Energy is transferred daily to and from the earth by the sun's radiation, rain, and wind. Each year, more than 6,000 times the amount of energy currently used by humans is striking the earth from the sun. In fact, less than 4% of the stored energy in the earth's crust comes from its hot molten center. Heat pumps use this energy stored in the earth's crust for heating. In the summer months, because the temperature of the earth's crust is cooler than the air just above it, summer air-conditioning heat loads can be transferred to the earth.

geothermal heat pumps
地源热泵

Geothermal heat pumps can pump heat in two directions. Because of this, they are normally used for space conditioning: heating and cooling. The same four basic components of heat pumps mentioned as "Air Source Heat Pumps"—the **compressor**, **condenser**, **evaporator**, and **metering device** — operate and control geothermal heat pumps. The four-way valve also controls the direction of heat flow in geothermal heat pumps.

9.1 Classification of Geothermal Heat Pumps

地源热泵分类

Geothermal heat pumps are classified as either **open-loop** or **closed-loop** systems. Open-loop, or water-source, systems use water from the earth as the heat transfer medium and then expel the water back to the earth in some manner. This process usually involves a well, lake, or pond. Open-loop systems need a large volume of clean water to operate properly. This same water supply also can be used for drinking and cooking.

Closed-loop, or earth-coupled, systems use a heat-transfer fluid, which is reused and circulated in plastic pipes buried within the earth or within a lake or pond. Closed-loop or earth-coupled systems are used where the water is rich in minerals, where local codes prohibit open-loop systems, or where not enough water exists to support an open-loop well water system. Both

compressor
压缩机

condenser
冷凝器

evaporator
蒸发器

metering device
节流装置

open-loop
开式系统

closed-loop
闭环系统

open-loop and closed-loop systems will be covered in detail in the following sections.

Whether the geothermal heat pump system is an open-loop, closed-loop, or earth-coupled system, water is still the source of heat when the system is operating in the heating mode. For this reason, these heat pumps are referred to as water-source heat pumps. Water-source heat pumps can be classified as either water-to-air or water-to-water. Water-to-air heat pumps are used to heat air in the occupied space; water-to-water heat pumps are used to heat water, which would be appropriate for an application such as a radiant heating system.

9.2 Open Loop Systems 开循环系统

Open-loop, water-source heat pump systems involve the transfer of heat between a water source and the air or water being circulated to a conditioned space. Remember, open-loop systems use water from the earth as the heat-transfer medium and then expel the water back to the earth. During the heating mode, heat is being transferred from the water source to the conditioned space. During the cooling mode, the heat removed from the conditioned space is deposited into the water. Fig. 9.1 illustrates both a heating and a cooling application of an open-loop, water-to-air heat pump. Defrost systems are not needed in geothermal heat pump systems. In the heating mode, water is supplied from a water source to a coiled, coaxial heat exchanger (Fig.9.2), by a circulating pump (Fig.9.1[a]). The heat exchange takes place between the water and the refrigerant. Refrigerant is carried in the outer section of the coaxial heat exchanger and the water flows in the inner tube (Fig.9.2[a]). Notice that the inner tube is ribbed to increase both the surface area and the heat-transfer rate between the fluids (Fig.9.2[b]).

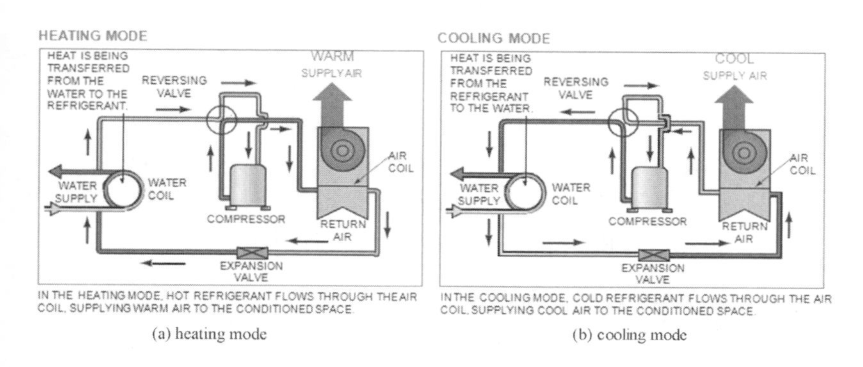

(a) heating mode (b) cooling mode

Fig. 9.1 An open-loop, water-source heat pump in both heating and cooling mode.

(a) Refrigerant flows in the outer tube while water flows in the inner tube.

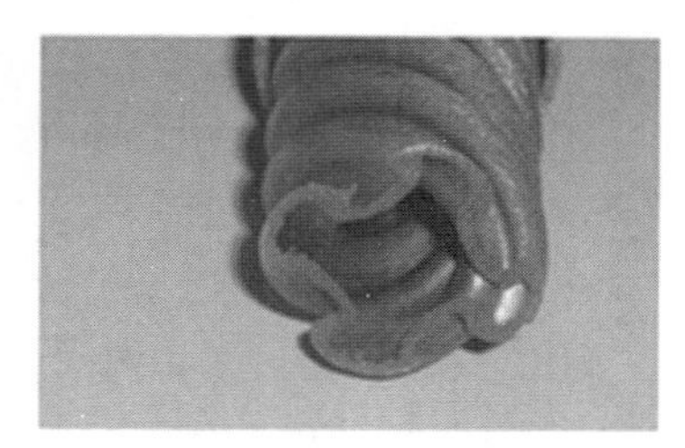

(b) Notice the shape of the inner tube. This increases the surface area and increases the rate of heat transfer.

Fig. 9.2 Cutaway view of a coaxial (tube-within-a-tube) heat exchanger.

The refrigerant side of the heat exchanger is the refrigeration system's evaporator. Heat is absorbed from the water into the vaporizing refrigerant. The refrigerant vapor then travels through the reversing valve to the compressor, where it is compressed. The heat-laden hot gas from the compressor then travels to the condenser. The condenser is a refrigerant-to-air, finned-tube heat exchanger located in the ductwork. It is often referred to as the air coil. Heat is then rejected to the air as the refrigerant condenses. A fan delivers the heated air to the conditioned space. The condensed liquid then travels through the expansion valve and vaporizes in the evaporator, absorbing heat from the water source. The process is then repeated. High-resistance, electric strip heaters can be used for auxiliary and/or emergency heat when the heat pump needs assistance.

In the cooling mode, the water loop acts as the condensing medium for the refrigerant (Fig.9.1[b]). Discharge gas from the compressor travels through the reversing valve to the outer portion of the coaxial (tube-within-a-tube) heat exchanger of the water loop. The refrigerant side of the heat exchanger is the refrigerant system's condenser. A refrigerant-to-water heat exchange takes place. The water loop absorbs heat from the refrigerant and condenses it. **Subcooled** liquid refrigerant then travels to the expansion valve and on to the air coil, where it evaporates and absorbs heat from the air. The air is cooled

subcooled
过冷的

and dehumidified. A fan delivers this air to the conditioned space. The heat-laden, superheated, refrigerant gas then travels from the air coil and through the reversing valve to the compressor, where the refrigerant is compressed. The process is then repeated.

For smaller residences, a single water-source heat pump is normally used, but larger houses may require more than one. For commercial applications, multiple heat pumps are combined into a system with a common water-piping loop. This loop provides a means for transferring the rejected heat from one heat pump in the cooling mode to another heat pump, which is in the heating mode. Commercial applications may also be connected to a well, a pond, a lake, or an earth-coupled closed-loop system. These systems may also contain a boiler and cooling tower to maintain desired loop temperatures (Fig.9.3).

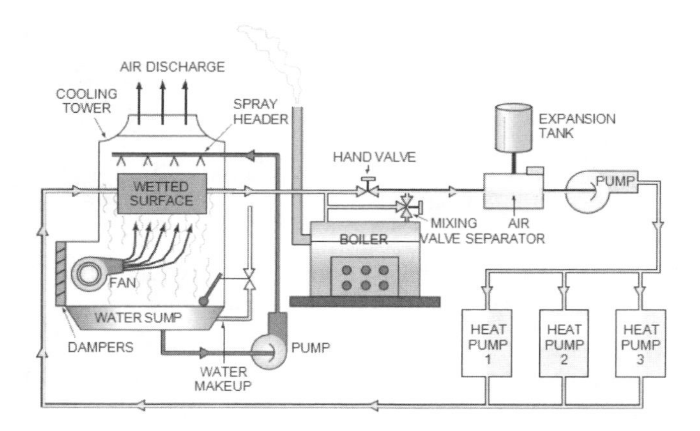

Fig. 9.3 An open-loop, water-source heat pump with boiler and cooling tower to maintain the loop temperature.

9.3 Closed Loop Systems 闭循环系统

In closed-loop heat pump systems, many yards of plastic pipe are buried in the earth. The pipe can be placed in either a horizontal or vertical configuration, depending on how much land is available and the soil composition. The loops of piping are called either ground loops or water loops, from Fig. 9.4 to Fig. 9.6. A completely sealed and pressurized loop of water, or water and anti-freeze solution, is circulated by a low-wattage centrifugal pump through the pipe buried in the ground. In the winter or heating mode, heat is transferred from the ground, through the plastic pipe, to the liquid in the ground loop. During the summer or cooling mode, heat is rejected away from the circulating fluid, through the plastic pipe, and into the ground.

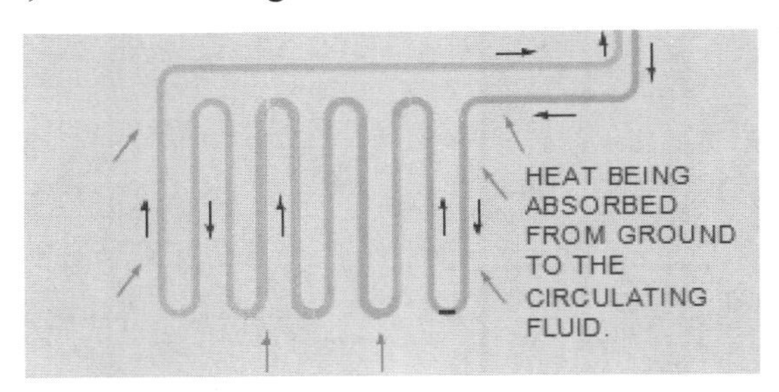

Fig. 9.4 A ground loop showing a series-vertical configuration in the heating mode.

Vertical systems are used when there is a shortage of land or space restrictions (Fig.9.4). If the soil is rocky, a rock bit is used on the rotary drill. If the land available is without hard rock, a horizontal loop should be considered (Fig. 9.5).

Fig. 9.5 A single-layer, horizontal ground loop in the heating mode.

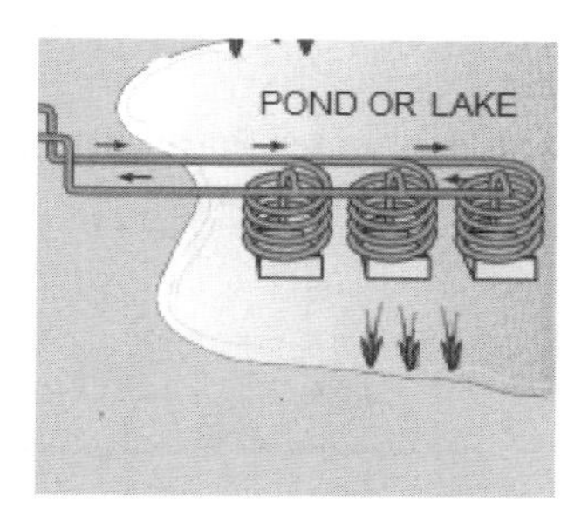

Fig. 9.6 A pond or lake loop.

An air loop is used to distribute heated or cooled air to the building. This is accomplished through a finned-coil, air-to-refrigerant heat exchanger located in the ductwork. A squirrel cage blower is used to move the air through the air distribution system. Fig. 9.7 and Fig. 9.8 illustrate the ground, refrigerant, and air loops in the heating and cooling modes, respectively.

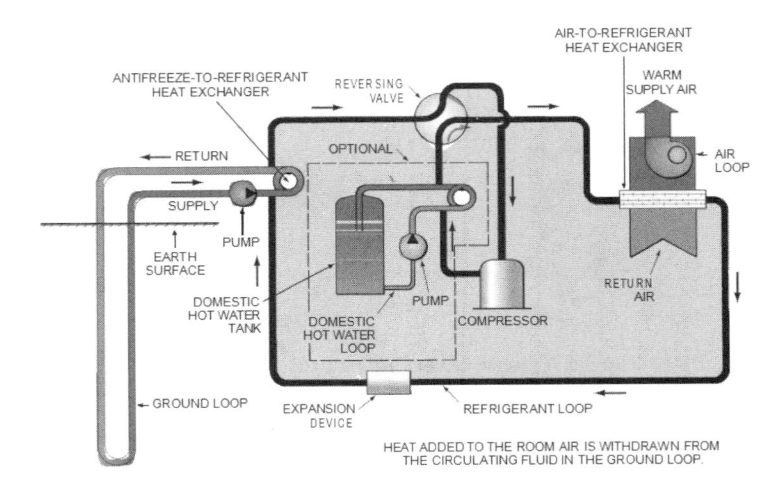

Fig. 9.7 A closed-loop, water-source heat pump in the cooling mode.

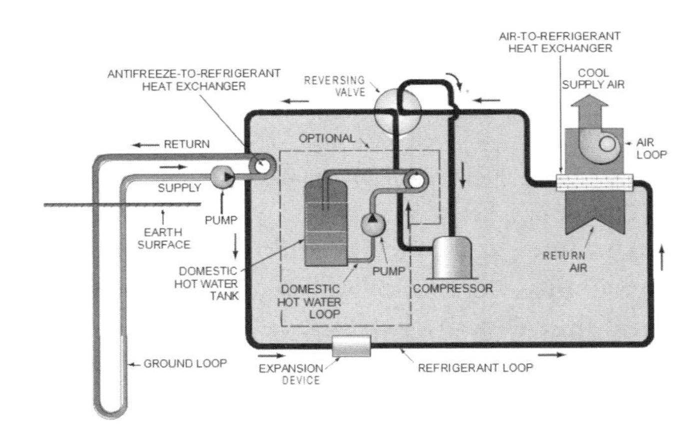

Fig. 9.8 A closed-loop, water-source heat pump in the heating mode.

本章重点内容介绍

地源热泵是制冷机，并且它的原理和空气源热泵相似，都是将热量从一个地方转移到另一个地方。

然而地源热泵用的是土壤或者地下水作为热源和散热器。地源热泵可以向两个方向输送能量，与“空气源热泵”一样，地源热泵也由四个基本部分组成，分别是压缩机、冷凝器、蒸发器和节流装置。

地源热泵可以分为开环和闭环两个系统。开环系统利用来自地球的水作为传热介质，然后以某种方式将水排回地球，开环系统需要大量清水才能正常运行。闭环系统使用一种传热流体，这种流体可以埋在地下或湖中或池塘中的塑料管中循环使用。在水富含矿物、当地法规禁止开环系统或水不足以支持开环井水系统的地方，使用闭环系统。无论地源热泵系统是开环还是闭环系统，当系统在供热模式下运行时，水仍然是热源。

开环水源热泵系统涉及水源与空气或水之间的热量传递，这些空气或水被循环到一定的空间。在加热模式下，热量从水源转移到有条件的空间。在冷却模式下，从调节空间排出的热量沉积到水中。工作方式是这样的，制冷剂蒸气通过换向阀流向压缩机，在那里被压缩。然后，来自压缩机的充满热量的热气体进入冷凝器。当制冷剂冷凝时，热量被排出到空气中，冷凝液随后通过膨胀阀在蒸发器中蒸发，从水源中吸收热量。然后重复该过程。在冷却模式下，水回路充当制冷剂的冷凝介质，从压缩机排出的气体通过换向阀流向水回路的同轴热交换器的外部。制冷剂与水进行热交换。水循环从制冷剂中吸收热量并使其冷凝。过冷的液态制冷剂随后进入膨胀阀并进入空气盘管，在那里蒸发并从空气中吸收热量。空气被冷却和除湿。充满热量、过热的制冷剂气体从空气盘管经过换向阀进入压缩机，在那里制冷剂被压缩。这个过程循环往复。

对于商业应用，多个热泵被组合成一个具有公共水管回路的系统。在闭环系统中许多塑料管被埋在地下。根据可用土地和土壤成分的不同，管道可以水平或垂直布置。一个完全密封和加压的水或水和防冻液构成的回路，通过一个低功率离心泵驱动制冷剂在管道循环。当缺乏土地或空间限制时，使用垂直系统。如果可用的土地没有坚硬的岩石，则应考虑水平环路。空气回路把加热或冷却的空气分配给建筑物。这是通过位于管道系统中的肋片、空气-制冷剂热交换器实现的。

Chapter 10

Chilled-Water Systems

冷冻水系统

Chilled-water systems are used for larger central air conditioning applications because of the ease with which chilled water can be circulated in the system. If refrigerants were piped to all floors of a multistory building, there would be too many possibilities for leaks, not to mention the expense of the refrigerant to charge such a large system. The design temperature for boiling refrigerant in a coil used for cooling air is 4.4 °C. If water can be cooled to approximately the same temperature, it can also be used to cool or condition air (Fig.10.1). This is the logic behind circulating chilled-water systems. Water is cooled to about 7.2 °C and circulated throughout the building to air-heat exchange coils that **absorb heat** from the building air. Water used in this way is called a secondary refrigerant. Water is much less expensive to circulate than refrigerant (Fig.10.2).

冷冻水系统包括冷冻水循环系统和热水循环系统。冷冻水循环系统是中央空调设备的冷冻水回水经集水器、除污器、循环水泵进入冷水机组蒸发器内，吸收了制冷剂蒸发的冷量，使其温度降低成为冷水，冷水进入分水器后再送入空调设备的表冷器或冷却盘管内，与被处理的空气进行热交换后，再回到冷水机组内进行循环再处理。

chilled-water
冷冻水

absorb heat
吸收热量

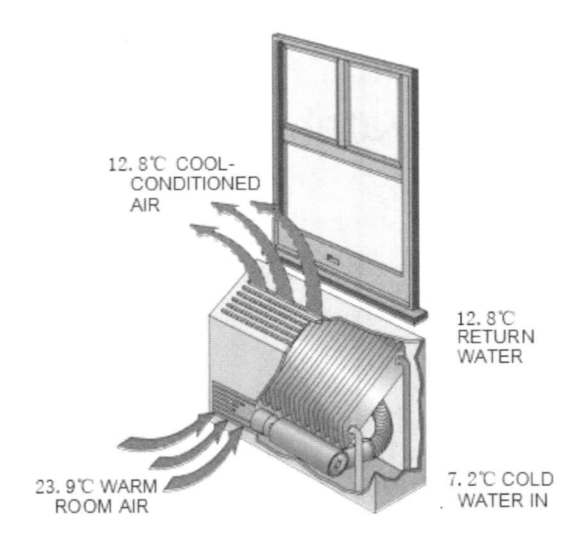

Fig. 10.1 Water circulating in a fan coil unit for cooling a room.

Chilled water circulated through a building is **typically** 7.2 ℃. Lower water temperatures may be used to make the building side of the chilled-water system more efficient. Many systems chill water to 5.6 ℃, and with the addition of electronic controls, some **manufacturers** are actually able to furnish 1.1 ℃ chilled water. The colder the circulating water, the smaller the terminal equipment—the coils and fans—can be. Also, with colder water, the interconnecting **piping** for the entire system can be smaller, which saves on piping supplies, fittings, and insulation for the entire system as well as on labor. Also, smaller fans and pumps save on electricity, because they consume less power. So the motor is stressed when the compressor is started up. In most

typically
代表性的，典型的

manufacturers
制造商

piping
管道

comfort-cooling installations, the load varies during the season as well as during the day. We know that when a system is designed for very high temperature conditions, the equipment will run at full load only about 3% of the time. The rest of the time it will run at reduced load. At times, the equipment will need to operate at 40% or 50% of its **capacity**; other times that it will need to run at only 15% or 20% of its capacity in mild weather.

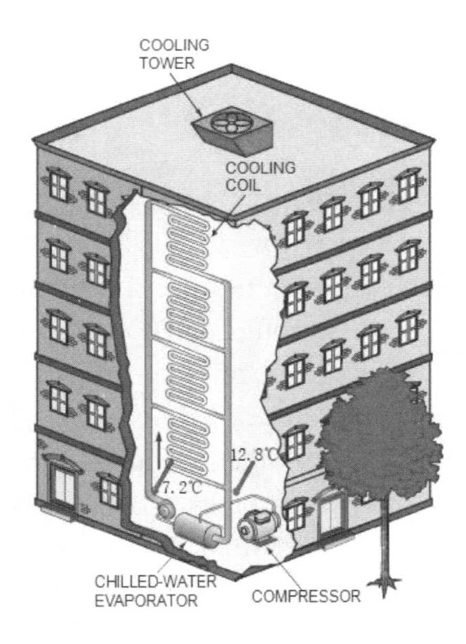

Fig. 10.2 Water is circulating throughout the building to the fan coil units. It is considered to be a secondary refrigerant.

Manufacturers have developed many capacity-control methods for their equipment. Some

capacity
能力，容量

chillers manager reduced capacity with variable-speed, or unloading, compressors. Variable-capacity control for each type of chiller will be discussed in the appropriate sections of this unit. Manufacturers usually rate chiller capacities and efficiencies at full-load operation. With energy costs rising and the increased emphasis on system efficiency and reduced environmental impact, manufacturers have implemented technologies to achieve efficient operation at part load. In the past, equipment did not typically operate as efficiently at part load.

10.1 Chillers
冷水机组

冷水机组又称为冷冻机、制冷机组、冰水机组、冷却设备等。因各行各业的使用比较广泛，所以对冷水机组的要求也不一样。

A chiller refrigerates circulating water. As the water passes through the evaporator section of the machine, often referred to as the chiller barrel, the temperature of the water is lowered. This chilled water is then circulated throughout the building, where it picks up heat. The typical design temperatures for a circulating chilled-water system are 7.2 ℃ for water furnished to the building and 12.7 ℃ for water returned to the chiller from the building, so the building heat adds about 5.5 ℃ to the water supplied by the chiller. Upon returning to the chiller, the heat is removed from the water, which is then recirculated through the building.

Cooling building is not the only application for water chillers, although most of this unit will cover comfort-cooling chillers — a thorough discussion of all chiller applications would be beyond the scope of this text. Many industries use chillers for process cooling. Some chillers are designed to circulate water and **glycol** (antifreeze) mixtures at temperatures well below freezing. These are used, for example, in the plastic-molding manufacturing business. Milk bottles are manufactured by the process of injecting hot plastic into a mold, and the mold must be cooled to solidify the plastic. When a water valve is opened at the correct time in the process, the plastic milk bottle is water-cooled slightly until solid and is then ejected from the mold. The flow of chilled water is stopped and the mold is ready to make another bottle. The chilled water allows this process to happen very quickly.

The textile industry is another example of process cooling. Mills use huge amounts of chilled water to maintain the conditions that allow thread to move fast through the machines. If the temperature and humidity are held at the correct levels, maximum production can be accomplished. In the years before chilled water was used, mills could only operate at full capacity about 3 months of the year. With chilled water, ideal conditions can be maintained 24 hours a day for 365 days a year. Many mills run 24 hours a day year-round, allowing one week annually for

glycol
乙二醇

maintenance.

There are two basic categories of chillers: the compression cycle chiller and the **absorption chiller**. The compression chiller uses a compressor to provide the pressure differences inside the chiller to boil and condense refrigerant. The absorption chiller uses a salt solution and water to accomplish the same result. These chillers are very different and are discussed **separately**. Manufacturers of both types provide different design features, but all of them attempt to build a reliable, low-cost chiller that will last a long time.

10.2 Compression Cycle in Chillers 冷水机组压缩循环

The compression cycle chiller has the same four basic components as a typical air conditioner: a compressor, an evaporator, a condenser, and a metering device. These components are generally larger, however, in order to handle more refrigerants, and they may use a refrigerant different from that used in air conditioners.

The heart of the compression cycle refrigeration system is the compressor. There are several types of compressor. Those common in water chillers are the reciprocating, scroll, screw, and centrifugal compressors. The compressor is the component in the system that both lowers the suction pressure and increases the discharge pressure. It can be

absorption chiller
吸收式制冷机

separately
分别地，单独地

thought of as a vapor pump. The compressor lowers the evaporator pressure to the desired boiling point of the refrigerant, which is about 3.3 °C for a chiller. It then builds the pressure in the condenser to the point that vapor will condense to a liquid for reuse in the evaporator. The typical condensing temperature for a chilled-water system equipped with a water-cooled condenser is about 40.5 °C. The technician can use temperatures to **determine** whether a typical chiller is operating within design parameters for pumping and compressing the vapor to meet the needs of a particular installation.

Compression cycle chillers may be classified as either high- pressure or low- pressure systems. Following is a discussion of high- pressure chilled water systems.

10.3 Reciprocating Compressors in High-Pressure Chillers
高压冷水机的往复式压缩机

Large reciprocating compressors in water chillers operate in the same way as those used in any other application, with a few exceptions. Reciprocating compressors range in size from about 1/2 hp to approximately 150 hp, depending on the application. Many manufacturers use multiple smaller compressors instead of a single, larger pump. Reciprocating compressors are

determine
判断

classified as positive displacement pumps and cannot pump liquid refrigerant without risk of damage. Several refrigerants have been used in reciprocating compressor chillers, most commonly R500, R502, R12, R134a, and R22. Because of ozone depletion and global warming concerns, environmentally friendly alternative refrigerants like R-134a and certain HFC refrigerant blends are now favored.

Large reciprocating compressors have many **cylinders** in order to produce the pumping capacity needed to move large amounts of refrigerant. Some of these compressors have as many as 16 cylinders; thus the machine becomes one with many moving parts and much internal friction. If one cylinder of the compressor fails, the whole system will go off-line. If one compressor fails in a multiple-compressor system, the others can carry the load, so multiple compressors give some protection from total failure. This is the reason, as well as for capacity control, that many manufacturers choose to use multiple smaller compressors in their systems.

All large chillers must have some means for controlling capacity or the compressor will cycle on and off. This is not satisfactory because most compressor wear occurs during start-up before adequate oil pressure is established. A better design approach is to keep the compressor running but operate it at reduced capacity.

cylinder
气缸

Reduced-capacity operation also smoothes out temperature fluctuations that occur from shutting off the compressor and waiting for the water to warm up to bring the compressor back on.

Reduced capacity for a reciprocating compressor is accomplished through cylinder unloading and variable frequency drives (VFDs). For example, suppose the chiller for a large office building uses a 100-ton compressor with eight cylinders and the chiller has 12.5 tons of capacity per cylinder. When all eight cylinders are pumping, the compressor has a capacity of 100 tons (8 × 12.5 = 100). As the cylinders are unloaded, the capacity is reduced. For example, the cylinders may unload in pairs, which would be 25 tons per unloading step. The compressor may have four unloading steps, which would give it four different capacities: 100 tons (eight cylinders pumping), 75 tons (six cylinders pumping), 50 tons (four cylinders pumping), and 25 tons (two cylinders pumping).

In the morning when the system first starts, the building may need only 25 tons of cooling. As the temperature outside rises and as people begin to enter the building, the chiller may need more capacity and the compressor will automatically load two more cylinders for 50 tons of capacity. As the heat load on the structure continues to rise, the compressor can load up to 100% capacity, or 100 tons. If the building — a hotel, for example, stays open at night, the compressor will start to unload as the outside temperature drops. It will unload down to 25 tons; if this is too much capacity, the chiller will then shut off. When the chiller is restarted, it will start up at the reduced capacity, lowering the starting current. A compressor cannot be unloaded to 0 pumping capacity because it would not be able to move any refrigerant through the system to return the oil in the system. Usually compressors will unload down to 25% to 50% of their full-load pumping capacity.

A big advantage of cylinder unloading is that the power needed to operate the compressor is reduced as capacity is reduced. The reduction in power **consumption** is not in direct proportion to compressor capacity, but the power consumption is greatly reduced at part load. In addition to the reduction in workload by cylinder unloading, the compression ratio is reduced; when a cylinder is unloaded, the suction pressure rises slightly and the head pressure is reduced slightly. Cylinder unloading is accomplished in several ways; blocked suction unloading and suction-valve-lift unloading are the most common.

10.3.1 Blocked Suction Unloading 堵塞的吸料器

consumption
消耗

Blocked suction is accomplished by placing

a solenoid valve in the suction passage to the cylinder being unloaded (Fig.10.3). If the refrigerant gas cannot reach the cylinder, no gas can be pumped. If a compressor has four cylinders and the suction gas is blocked to one of them, the capacity of the compressor is reduced by 25% and the compressor then pumps at 75% capacity. Power consumption also decreases by approximately 25%. Power consumption is related to the amperage draw of the compressor. Amperage is typically measured in the field by using a clamp-on ammeter. When a compressor is running at half capacity, the amperage will be about half the full-load amperage. (The power consumption of a compressor is actually measured in watts. Using amperage as a measure of compressor capacity is close enough for field troubleshooting.)

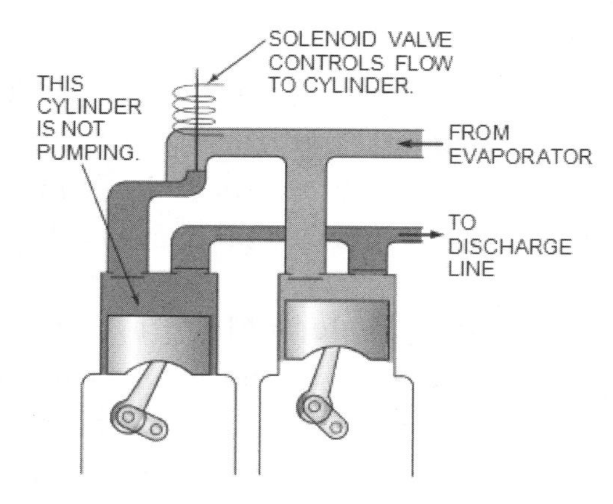

Fig. 10.3 Blocking the flow of refrigerant to one of the cylinders in a reciprocating compressor will unload that cylinder's load from the system.

10.3.2 Suction Valve Lift Unloading 吸气阀提升卸载

If the suction valve is lifted off the seat of a cylinder while the compressor is pumping, the cylinder will stop pumping. Gas that enters the cylinder will be pushed back out into the suction side of the system on the upstroke. There is very little resistance to the refrigerant being pumped back into the suction side, so it requires almost no energy. As in the example of blocked suction, power consumption will be reduced. One of the advantages of lifting the suction valve is that the gas that enters the cylinder will contain oil, and good cylinder lubrication will occur even while the cylinder is not pumping. Compressor unloading could be accomplished by letting hot gas back into the cylinder, but this is not practical because power output will not be reduced. When the gas has been pumped from the low-pressure side to the high-pressure side of the system, the work has been accomplished. Also, allowing hot gas back in would overheat the cylinder by making it compress the hot gas again.

Except for cylinder unloading, the large reciprocating chiller compressor is the same as smaller reciprocating compressors. Most compressors over 5 hp have pressure-lubricating systems. The lubricating pressure is provided by an oil pump that is typically mounted on the end of the compressor shaft and driven by the shaft (Fig.10.4). The oil pump picks up oil in the sump at evaporator pressure and delivers it to the **bearings** at about 206.8 to 413.7 kPa greater than suction pressure, called net oil pressure (Fig.10.5). The compressors also have an oil safety shutdown in case of oil pressure failure, which has a time delay of about 90 sec to allow the compressor to get started and to **establish** oil pressure before it shuts the compressor off.

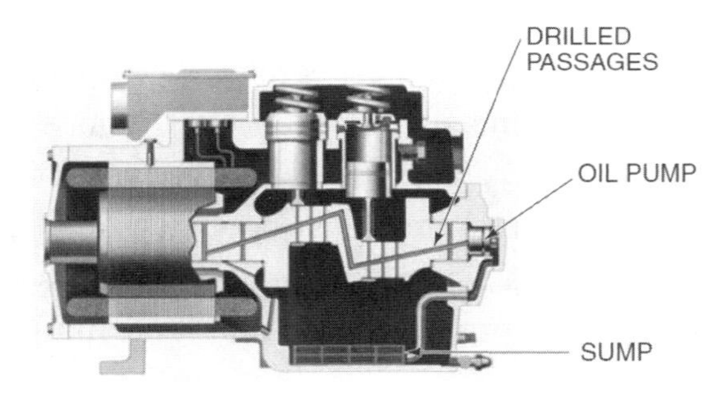

Fig. 10.4 The pressure lubrication system for a reciprocating compressor. Notice that the oil pump is on the end of the shaft. The oil is picked up from the sump and pumped through the drilled passages to all moving parts.

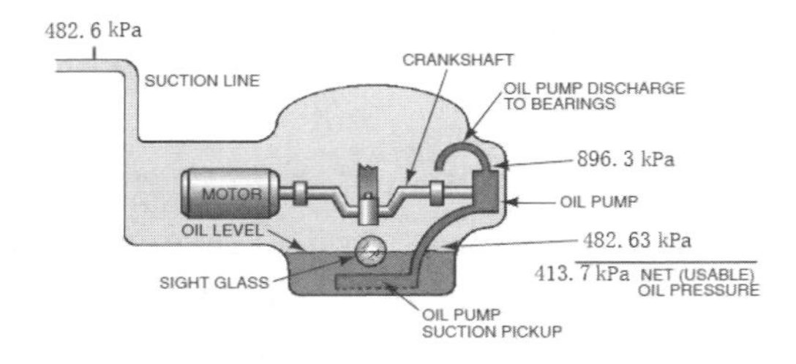

Fig. 10.5 The oil pump's inlet pressure is the same as the system's suction pressure in the crankcase. The oil pump then increases the oil pressure to the pump's discharge pressure.

bearings
轴承

establish
建立，设立

10.3.3 Variable Frequency Drives 变频驱动器

Variable frequency drives (VFDs), electronically alter the frequency of alternating-current (AC) sine waves through a device called an inverter. Today, compressor motors can operate within a wide range of speeds that can vary capacity without shutting off. The speeds used are in conjunction with electronic expansion valves that also have an almost unlimited ability to vary refrigerant flow to the evaporator. The condenser water pumps and the chilled-water pumps may also have VFDs that can vary the water flow. The system flow rates must be synchronized by computer and take into account the desired system conditions to prevent problems and maintain minimum and maximum flow rates for all the devices that are being coordinated.

10.4 Rotary Screw Compressors in High Pressure Chillers 高压冷水机的旋转螺杆压缩机

Most of the major manufacturers are building rotary screw compressors for larger-capacity chillers, which use high-pressure refrigerants. The rotary screw compressor is capable of handling large volumes of refrigerant with few moving parts (Fig.10.6). A positive displacement compressor, this type of compressor is able to handle some liquid refrigerant without compressor damage, unlike the reciprocating compressor, which cannot handle liquid refrigerant. Rotary screw compressors are manufactured in sizes from about 50 to 700 tons of capacity and are reliable and trouble-free. Manufacturers build both semi-hermetic rotary screw compressors and open-drive compressors. The open-drive models are direct drive and must have a shaft seal to contain the refrigerant at the point where the rotating shaft penetrates the compressor shell. It is common practice to start up open-drive compressors regularly during the off-season to keep the shaft seal lubricated. Long periods of system downtime can cause the shaft seal to dry out and shrink somewhat, allowing refrigerant to leak from the system.

(a) Cutaway view of a rotary screw compressor

(b) Closeup of the nesting screws

(c) Closeup of the screws

Fig. 10.6 Rotary screw compressors.

Capacity control for a rotary screw compressor may be accomplished by means of a slide valve that blocks the suction gas before it enters the rotary screws in the compressor or by means of compressor VFD speed control. The slide valve is typically operated by differential pressure in the system. The slide valve may be moved to the completely unloaded position before shutdown so that on start-up, the compressor is unloaded, reducing the inrush current. Most of these compressors can function from about 10% to 100% load with sliding graduations because of the nature of the slide valve unloader. This is in contrast to the step capacity control of a reciprocating compressor, which unloads one or two cylinders at a time.

Screw compressors pump a great deal of oil while compressing refrigerant, so they typically have an oil separator to return as much oil to the compressor reservoir as possible (Fig. 10.7). The oil is moved to the rotating parts of the compressor by means of pressure differential within the compressor instead of an oil pump. Oil is also accumulated in the oil reservoirs and moved to the parts that need lubrication by means of gravity. The rotating screws are close together but do not touch. The gap between the screws is sealed by oil that is pumped into the rotary screws as they turn. This oil is separated from the hot gas in the discharge line and returned to the oil sump through an oil cooler. This separation is necessary because if too much oil reaches the system, a poor heat exchange will occur in the evaporator and cause loss of capacity. The natural place to separate the oil from the refrigerant is in the discharge line. Manufacturers use different methods for this process, but all use some means of oil separation.

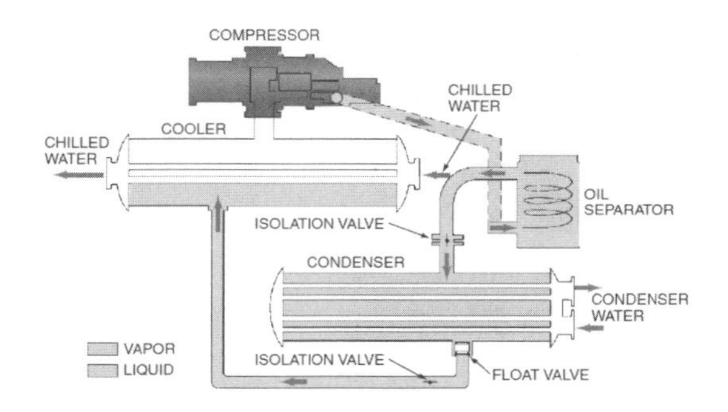

Fig. 10.7 An oil separator for a rotary screw compressor.

本章重点内容介绍

冷冻水系统应用于大型的中央空调，因以水作为冷媒（工作介质）在系统中进行循环，因此得名冷冻水系统。直接输送制冷剂会有一定的损失风险：当管路发生泄露时，即使不计制冷剂的腐蚀性以及毒性，也会损失大量的制冷剂。在冷冻水系统中，涉及几个基本的参数，例如，用于冷却冷冻水的制冷剂是4.4 ℃，而冷冻水被降温至 7.2 ℃即可进入循环，用于冷却空气。冷冻水系统要比制冷剂系统更加廉价，因为整个系统中管道占地更少，而水泵和风机所需动力也更小。

本章分别介绍了冷水机组、冷水机组的压缩循环，以及整个系统的核心部件：压缩机（往复式、离心式、涡旋式、螺杆式）。

冷水机组的核心由四个部分构成：蒸发器、冷凝器、压缩机和节流阀。而与制冷剂系统不同的是，冷水机组中的制冷剂只在这四个基本构件中循环，而真正承担提供用户冷量的是充当冷媒的循环水。在本章的内容中，还涉及一些关于冷冻水以及制冷剂的设计温度和实际工况问题。

往复式压缩机和螺杆式压缩机最大的区别就是往复式压缩机可以使多个压缩气缸同时工作，但同时也会出现一些问题，比如，如果一个气缸出现故障，整个压缩机设备都必须停止工作，但多级的压缩机可以不停机，而是由其他气缸分摊故障气缸的负载。在学习往复式压缩机的过程中，还会细化分析一些参数，例如变频问题。

螺杆式压缩机是目前最为普遍的大容量冷水机组所使用的容积式压缩机，其能用较少的运动部件处理大量的制冷剂。与往复式压缩机不同的一点是，螺杆式压缩机在润滑上的要求要更加严密。

Chapter 11

Phase Change Material

相变材料

In recent decades, energy demands have increased rapidly due to the increase of the growth rate of population and the living standard of people. Housing apartments and commercial buildings play an important role in energy-intensive consumers. Approximately 1/3 of all energy has been used for illumination, heating and air-conditioning of buildings.

相变材料是指在温度不变的情况下改变物质状态并能提供潜热的物质。转变物理性质的过程称为相变过程，这时相变材料将吸收或释放大量的潜热。这种材料在生活中被广泛应用，是节能环保的最佳绿色载体。

Thermal energy storage (TES) has been considered one of the most efficient ways for building to promote the energy utilization efficiency and reduce the energy consumption. TES is mainly classified as sensible thermal energy storage (STES) and latent thermal energy storage (LTES). STES stores/releases thermal energy by raising the temperature of a solid or a liquid, and it depends on the specific heat capacity of materials during the process of charging and discharging. LTES stores/releases thermal energy by solid to solid, solid to liquid, or liquid to gas, and it depends on the accumulated energy densities of materials at almost **isothermal** conditions. Moreover, LTES is widely used in building TES system due to its higher storage density than STES and nearly isothermal heat storage process.

PCMs have been employed for LTES in building for several decades. PCM is a substance that can absorb, store and release a large amount of thermal energy during phase change process at specific temperature. PCMs can not only save thermal energy to adjust the indoor temperature fluctuation, but also reduce the mismatch between energy supply and energy consumption.

11.1 Classification of PCMs 相变材料分类

Different classification methods of PCMs are listed in Tab. 11.1. The most common way to distinguish PCMs is by dividing them into organic, inorganic and eutectic based on the chemical composition.

Organic PCMs have the advantages of chemically stable, non-reactive, little or no super cooling and high heat of fusion. In general, organic PCMs are divided into **paraffin** and non-paraffin. Paraffin is a hydrocarbon that has the chemical structure C_nH_{2n+2}. The more carbon

isothermal 等温的

paraffin 煤油

atoms present in the paraffin, the higher the melting temperature. Paraffin is safe, reliable, predictable, inexpensive, non-corrosive and chemically stable. However, low thermal conductivity of about 0.2 W/(m·K) and liquid leakage during phase change process hinder their applications in some fields. Non-paraffin includes fatty acids and their fatty acids eaters, alcohols and glycols. Their properties are similar to paraffin, while the expense of non-paraffin is about three times more than paraffin.

Tab. 11.1 Classification of PCMs

Classification Method	Type
phase change type	solid-liquid PCMs
	solid-solid PCMs
	solid-gas PCMs
	liquid-gas PCMs
chemical composition	organic PCMs
	inorganic PCMs
	eutectic PCMs
phase change temperature	Low temperature PCMs($<$ 15 °C)
	Medium temperature PCMs(15 °C~90 °C)
	high temperature PCMs($>$ 90 °C)
energy storage	heat storage PCMs
	cold storage PCMs

Inorganic PCMs have the advantages of high thermal conductivity, **non-flammable**, ready availability and low cost. However, most of them are corrosive to most metals, undergo super cooling during solid-liquid transition and undergo phase decomposition. Most inorganic PCMs

non-flammable
不易燃的

are **hydrated** salts, which consist of an alloy of inorganic salts and water. In many cases, hydrated salts lead to the salt releasing water and turning into a salt in its anhydrous form or a different salt during the melting process. Moreover, some hydrated salts occur super cooling because of poor nucleating properties. This is a critical flaw for the melting temperature.

Eutectic PCMs consist of two or more components such as organic-organic, inorganic-inorganic and organic-inorganic, and each of the eutectics melts and freezes congruently. The main advantages of eutectic PCMs are the adjusted melting temperature by combining different weight percentages of components, high thermal conductivity and density, non-super cooling and non-segregation. However, eutectic PCMs have smaller latent and specific heat capacities than other PCMs.

11.2 Desirable Properties of PCMs in Building Wall 相变材料在建筑墙体中的理想特性

hydrated
使吸入水分

eutectic
易熔的

In order to be used for building wall, PCMs must meet certain desirable properties such as thermal-physical, chemical, kinetic, economic and environmental. The following conclusions are drawn:

(1) Suitable phase change temperature for building wall;

(2) High latent heat, high thermal conductivity and high specific heat;

(3) Small volume change during phase change progress;

(4) Non-toxic, non-flammable, non-degradable and non-corrosive to assure safety;

(5) Good compatible with construction and encapsulated materials;

(6) High rate of crystal growth and nucleation to avoid super cooling of PCMs;

(7) Low cost and commercially available;

(8) Low environmental impact during service life;

(9) Having recycling potential.

However, it is difficult to find a PCM which can completely satisfy all these desirable properties. In practice, the phase change temperature and the latent heat of PCMs are considered first, and other properties are considered. Tab.11.2 shows the thermo physical properties of some PCMs used in building.

Tab. 11.2 Thermo physical properties of some PCMs used in building

PCM	Melting temperature /°C	Heat of fusion /(J/g)
Paraffin	26.83	136.2
capric acid	33.03	154.42
lauric acid	43.32	179.85
heptadecane	22.23	156.8
n-octadecane	27.5	239.4
$CaCl_2$-$6H_2O$	29.8	191
$LiNO_3 \cdot 3H_2O$	30	296
76.3% lauric acid+ 23.7% stearic acid	38.99	155.94
50% $CaCl_2$+ 50% $MgCl_2 \cdot 6H_2O$	25	95

11.3 Incorporation of PCMs into Building Wall

相变材料与建筑墙体的融合

相变材料在建筑中的利用，将极大推进绿色建筑和节能环保产业的发展。

PCMs can be incorporated into building wall by direct incorporation, immersion, **encapsulation** and form-stable PCMs.

11.3.1 Direct Incorporation

直接融合

encapsulation 封装

gypsum 石膏

In the direct incorporation technique, PCMs are directly mixed with building materials such as **gypsum** and cement paste during production. It is the simplest, practicable and economical method. Feldman et al. prepared a energy storage gypsum wallboard by the direct incorporation of 21%-

22% commercial grade Butyl Stearate (BS). The physic-mechanical characteristic and the thermal storage property were investigated. Compared with the common gypsum wallboard, the energy storage gypsum wallboard had a ten-fold increase in capacity for the storage and discharge of heat. However, the leakage of PCMs after several thermal cycles may affect the mechanical and durability properties of construction elements.

11.3.2 Immersion
浸 没

In the immersion technique, construction elements such as concrete, brick block and wallboard are dipped into the liquid PCM and absorb PCM by capillary force. Hawes reported the immersion technique that the porous concrete was filled with liquid PCMs for the first time. The results showed that the effectiveness and immersion time for liquid PCM to be soaked into the porous concrete were mainly based on the absorption capacity of concrete, the melting temperature and the type of PCMs. In later work, Hawes determined the effects of different types of porous concrete on the absorptive of PCMs in concrete. However, the construction elements also may leak after large amounts of thermal cycles and be affected by the mechanical and durability properties.

11.3.3 Encapsulation
封　装

In the encapsulation technique, PCMs are encapsulated with suitable coatings or shell materials before used into construction elements (see Fig.11.1). Encapsulation can avoid direct contact between PCMs and surrounding environment, prevent the leakage of PCMs when it is in a liquid state and increase the efficient heat transfer.

Based on particle size, encapsulation can be classified into macroencapsulation, microencapsulation and nanoencapsulation.

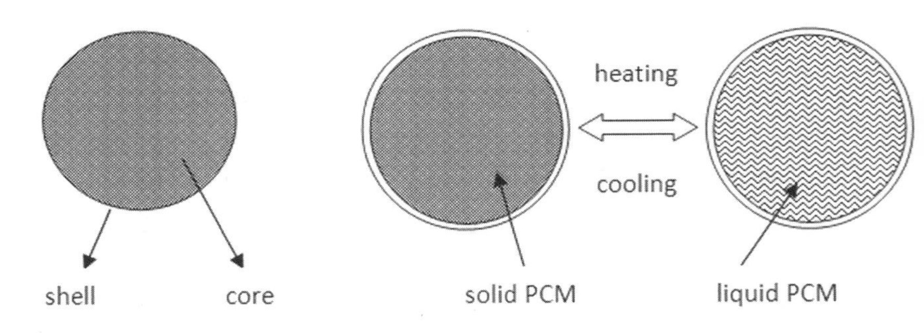

Fig. 11.1 Working principle of encapsulated PCMs.

Macroencapsulation refers to the PCMs encapsulated in a container such as tubes, spheres and panels for subsequent use in construction elements. The particle size of macroencapsulation is commonly larger than

1mm. Macroencapsulation is easier to ship and handle, improves compatibility of PCMs with the surrounding by acting as a barrier and reduces external volume change. However, low thermal conductivity of PCMs leads to a slower energy uptake and release which may prevent the products from discharging completely overnight. Therefore, the choice of the shell materials of macroencapsulation plays an important role in regulating properties.

Microencapsulation is a technique in which the particles ranging from 1μm to 1,000 μm. It comprised of the outer shell and PCM filled in centre. Microencapsulation can provide high heat transfer rate, prevent the leakage of PCMs during phase change process, improve thermal reliability and chemical stability, and control PCMs' volume changes. However, microencapsulation leads to higher heat transfer rates that result in rapid melting and solidification of encapsulated PCMs. Three methods (physical, chemical and physic-chemical) (Tab. 11.3) have been reported in the literature. At present, prominent studies are mainly focused on the preparation and performance of microencapsulated PCMs.

Nanoencapsulation is a technique which has the particles in the range of 0~1,000 nm. Nanoencapsulation maintains the advantages of microencapsulation, also improves the specific area and interaction force between interfaces of particles. Therefore, nanoencapsulation has became the hot topic in buildings for thermal energy storage system.

Tab. 11.3 Microencapsulation methods

Physical methods	Physic-chemical methods	Chemical methods
pan coating	Ionic gelation	interfacial polymerization
air-suspension	coating coacervation	suspension polymerization
centrifugal extrusion	sol-gel method	emulsion polymerization
vibrational nozzle		
spray drying		
Solvent evaporation		

11.3.4 Form-stable PCMs 形状稳定的相变材料

In the form-stable PCMs technique, porous materials such as high-density polyethylene, styrene, expanded perlite and expanded graphite are used to fabricate form-stable PCMs. These porous materials can not only provide good mechanical strength to whole form-stable PCM, but also prevent the leakage for PCMs during phase change process. Form-stable PCMs have advantages of large apparent specific heat, keeping the shape stabilized during phase change process and good thermal reliable over a long period. Therefore, it can be concluded that form-stable PCM is a promising technique for thermal energy storage system in buildings.

11.4 Applications of PCMs in Building Wall
相变材料在建筑墙体中的应用

Current researches mainly focus on integrating PCMs into building wall such as wallboard, concrete and mortar.

11.4.1 PCMs Used in Wallboard
墙板中使用的相变材料

Wallboard is suitable for incorporation of PCMs because of its low cost, large heat exchange area, small heat exchange depth and widespread used in lightweight construction. Sari et al. prepared a novel form-stable phase change wallboard by incorporating eutectic mixture of capric and stearic acid into gypsum wallboard. The maximum of 25% of eutectic PCMs could be incorporated into the gypsum wallboard, and the melting temperature and latent heat of the form-stable phase change wallboard were 24.68 °C and 48.32 J/g, respectively. After 5,000 thermal cycles, the melting and freezing temperature of the form-stable phase change wallboard showed no obvious changed. Moreover, the results were compared with ordinary wallboard. It was concluded that the form-stable phase change wallboard could reduce the indoor temperature by 1.3 °C and improve the indoor thermal comfort.

11.4.2 PCMs Used in Concrete 用于混凝土的相变材料

Concrete is the most commonly used building material for incorporation with PCMs due to its wide spread application, various shapes and sizes, large heat exchange area and small heat exchange depth. Fig. 11.2 shows the heating and cooling function of concrete wall incorporated with PCMs to maintain pleasant human comfort temperature in indoor room. During the daytime, concrete wall absorbs solar radiation heat by PCMs. When the indoor temperature is lower than the melting temperature of PCMs, the heat that stored in concrete wall is released into the room to maintain indoor thermal comfort and reduce the indoor temperature fluctuation.

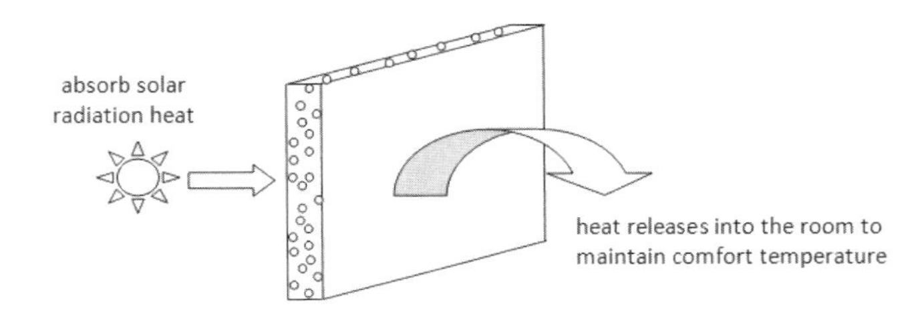

Fig. 11.2 Heating and cooling function of concrete wall incorporated with PCM to maintain pleasant human comfort temperature in indoor room

The effective thermal conductivity of novel form-stable basalt fiber composite concrete with PCMs was investigated through experimental and numerical method. Paraffin as PCM was

dispersed into the matrix at the liquid state, and the basalt fibers as additive were used to improve the elastic modulus and strength. The results showed that the effective thermal conductivity of concrete decreased with the increasing of porosity of paraffin or basalt fiber. The finite contact thermal resistance at the interface affected the composites at solid state. Moreover, the numerical model assisted in designing form-stable fiber concrete with PCMs due to the numerical results in accord with data.

11.4.3 PCMs Used in Mortar 砂浆中使用的相变材料

PCMs also have been incorporated into mortar to adjust the indoor temperature and reduce the energy consumption. Sandra et al. investigated the physical and mechanical properties of mortar based on different binders and different contents of PCMs. The experimental tests showed that the incorporation of PCMs caused an increase in the amount of water added to the mortar and this increase in the water/binder ratio was related to the fineness characteristics of PCMs. The flexural strength, compressive strength and adhesion of mortar decreased with the content of PCMs. Based on these results, the incorporation of PCMs in mortars had an adequate workability and a similar aspect to traditional mortar.

dispersed
分散的

The fire behavior of cement mortar with different mass fractions of PCMs was investigated by Haurie et al. The PCM GR27 was embedded into mortar replacing part of lightweight aggregate, and the PCM enthalpy-temperature curve was measured using the T-history method. The experimental results showed that the addition of PCMs in mortar caused a reduction of thermal diffusivity. 10% GR27 limited smoke release, as well as low values of heat release capacity. While 25% GR27 did not significantly change the performance of mortar.

本章重点内容介绍

近几十年来，由于人口增长率和人们生活水平的提高，能源需求迅速增加。潜热储能系统通过固体储存 / 释放热能到固体，固体储存 / 释放热能到液体，或液体储存 / 释放热能到气体，这取决于材料在等温条件下的累积能量密度。此外，由于比显热储能系统具有更高的存储密度和近等温储热过程，潜热储能系统在储热系统的构建中得到了广泛的应用。

有机相变材料具有化学稳定、无反应、微冷或无超冷和高聚变热等优点。一般来说，有机相变材料分为石蜡和非石蜡。石蜡是一种碳氢化合物，石蜡中碳原子越多，熔化温度越高。石蜡安全、可靠、可预测、廉价、无腐蚀性，具有化学稳定性。然而，约 0.2 W/(m · K) 的低热导率和相变过程中的液体泄漏阻碍了它们在某些领域的应用。非石蜡包括脂肪酸、醇和糖等。它们的性质类似于石蜡。而非石蜡的价格大约是石蜡的三倍。

无机相变材料具有导热系数高、不易燃、易得、成本低等优点。然而，它们对大多数金属具有腐蚀性，在固液转变过程中经历超冷，并经历相分解。大多数无机相变材料是水合盐，由无机盐和水的合金组成。在许多情况下，水合盐导致盐释放水，并在熔化过程中变成盐。此外，一些水合盐由于成核性能差而发生超冷。这是熔化温度的一个关键缺陷。

共晶相变材料由有机－有机、无机－无机和有机－无机两种或两种以上的成分组成，每种共晶体均一致地熔化和冻结。共晶相变材料的主要优点是通过组合不同重量百分比的组分、高导热率和密度、非超冷和非偏析来调节熔化温度。然而，共晶相变材料比其他相变材料具有更小的潜热和比热容量。

为了用于建筑墙体，相变材料必须满足某些理想的性能，如热物理、化学、动力学、经济和环境的要求。然而，很难找到一个能够完全满足所有这些理想属性的相变材料。在实践中，首先考虑相变材料的相变温度和潜热，然后考虑其他性质。

可在建筑墙体中直接掺入、浸泡、封装形式稳定的相变材料。其中技术之一是微胶囊，其粒子直径从 1 μm 到 1 000 μm。它由外壳和填充在中心的相变材料组成。微胶囊可以提供较高的传热速率，防止相变过程中相变材料的泄漏，提高热可靠性和化学稳定性，控制相变材料的体积变化。然而，微胶囊化导致更高的传热速率，导致封装的相变材料快速熔化和凝固。目前，突出的研究主要集中在微胶囊化相变材料的制备和性能方面。

墙板因其成本低、热交换面积大、热交换深度小而广泛应用于轻型建筑中，适合于相变材料的掺入。通过在石膏墙板中加入癸酸和硬脂酸的共晶混合物，制备了一种新型的稳定相变墙板。共晶相变材料可掺入石膏墙板，形式稳定相变墙板的熔化温度和潜热分别为 24.68 ℃和 48.32 J/g。经过 5 000 次热循环后，稳定相变墙板的熔化温度和冻结温度没有明显变化。并与普通墙板进行了比较，结果表明，稳定相变墙板可以降低室内温度 1.3 ℃，能提高室内热舒适性。混凝土是最常用的与相变材料结合的建筑材料，因便于制作成各种形状和尺寸，具有较大的热交换面积和较小的热交换深度，所以应用广泛。在白天，混凝土墙通过相变材料吸收太阳辐射热量。当室内温度低于相变材料的熔化温度时，将储存在混凝土墙体中的热量释放到室内，以保持室内热舒适性，减少室内温度波动。

在形式稳定的相变材料技术中，采用高密度聚乙烯、苯乙烯、膨胀珍珠岩和膨胀石墨等多孔材料制备形式稳定的相变材料。 这些多孔材料不仅能为整体稳定的相变材料提供良好的机械强度，而且能防止相变过程中相变材料的泄漏。稳定相变材料具有表观比热大、相变过程中形状稳定、热可靠等优点。因此，可以得出结论，形式稳定的相变材料是一种很有前途的建筑储能系统技术。同时纳米封装技术是一种粒子在0~1 000 nm的技术。纳米封装保持了微胶囊的优点，也提高了粒子界面之间的比面积和相互作用力。因此，纳米封装已成为建筑热能存储系统的研究热点。

Chapter 12
Thermal Comfort
热舒适性

12.1 Comfort and Discomfort
舒适和不适

One of the goals of the environmental engineer and architect is to ensure comfortable conditions in a building. Thermal pleasure can only be achieved locally over part of the body or temporarily in the context of a situation which is in itself uncomfortable. Continuous thermal pleasure extending over a period of hours is not possible. We are left simply with the idea of comfort as a lack of discomfort: this may seem an uninspiring definition, but nevertheless it presents a real practical challenge.

热舒适是人对周围热环境所做的主观满意度评价。主要分为三个方面：物理方面、生理方面、心理方面。评价指标主要有贝氏标度、ASHRAE 标度和 ISO 标准。

General thermal discomfort will be felt if a person is either too hot or too cold. In addition there are several potential sources of local discomfort, such as cold feet or **draughts**.

draught
穿堂风

Any guide to comfort must relate these forms of discomfort to the physical variables of the environment, so that a permissible range of the variables may be recommended. It is conventional to treat overall thermal discomfort in terms of thermal sensation.

For other forms of discomfort it is not possible to base the definition of discomfort simply on a scale of overall thermal sensation. It is possible to be thermally neutral and the temperature should be neither raised nor lowered, yet still be uncomfortable because of some non-uniformity in the environment, such as a draught or radiant **asymmetry**. The most direct way of finding if some is uncomfortable is simply to ask him. Usually the subject is asked to make a decision as to whether he is comfortable or uncomfortable, or whether he finds the thermal conditions acceptable or uncomfortable. Different people may be expected to become uncomfortable at different levels of external stress, so if the proportion of people voting uncomfortable is plotted against the level of stimulus, we should expect to get a curve of the shape. This is an ideal presentation, since the end user of the information is now able to trade off the proportion of people made uncomfortable against the cost of controlling the uncomfortable stimulus.

asymmetry
不对称

How well can we achieve such a presentation in practice? In laboratory studies the control and measurement of the potentially comfortable stimulus can be done to any required degree of accuracy. The stimulus might be, for instance, radiant asymmetry. A subject is then exposed to several levels of radiant asymmetry, with the other environmental withheld constant; the subject is asked to say if he is uncomfortable or not. After sufficient subjects have been exposed to various levels of radiant energy, a curve of the form may be plotted.

This experimental **paradigm** has formed the basis for much of the work on comfort, but has some weaknesses. The subject is asked to make a decision about whether he is uncomfortable or not; he has to make this decision about whether he is uncomfortable or not; he has to make this decision in the unfamiliar surroundings of a laboratory after a limited exposure time. The results may then be applied to a population at its normal place of work. This ignores the fact that a person's judgment may be very dependent on context. For instance, the radiant asymmetry provided by direct sunshine is far higher than the level which is regarded as satisfactory from a radiant heating system, yet people seek out and welcome sunshine. While complaints of draughts are common in modern offices, a gentle walking pace produces an air speed over the body of greater than 1 m/s, which would normally be regarded as very draughty, yet no one complains of draught discomfort while walking. The

paradigm
范例

response to a stimulus depends on the general surroundings and expectation of the person.

A particular example is the range effect. When an observer experiences a range of stimuli and is asked to rate them on a category scale, he tends to rate them by putting the stimuli from the middle of the range into the central categories of the rating scale. This has been clearly demonstrated in experiments on the acceptability of noise. Subjects exposed to a range of sound levels tend to place the boundary between acceptable and unacceptable noise at the center of the range of noise which they have been exposed. Thus people who are exposed to high noise levels will apparently tolerate more noise than people who have only been exposed to low noise levels. The range effects do not apply only to the range of stimuli provided by the experimenter. People carry their own standards with them, based on their general experience, with which they compare a new stimulus. Thus the meaning of the words comfortable or uncomfortable will not have an absolute value, but will be relative to his experience and expectation. When Gagge et al. put young men in an environmental **chamber** at 40 °C, the subjects rated the environment as slightly uncomfortable. There is little doubt that the equivalent environment in an office would be regarded as intolerable. However, the subjects knew they were in a physiological laboratory, were expecting to sweat, and did not object

chamber
会议厅

strongly to the experience.

It is clear, however, that their comfort rating cannot be transferred to a different situation. The standards of acceptability are set by the range of stimuli that people are used to. Poulton points out that this implies that the goal of providing a universally acceptable environment may be ever **receding**. If people do make judgments of acceptability on the basis of their own experience, then the maximum acceptable level will fall as the general level falls. If the noise level in a district is reduced, the level at which a noise becomes unacceptable will also be reduced and so the loudest noise will always be too loud. Air conditioning engineers are often heard to complain that standards of expectation rise as fast as the standard of air conditioning, so that the level of complaints stays constant.

12.2 Thermal Comfort in Artificial Environments 人工环境中的热舒适性

Modern humans demand high levels of thermal comfort in artificial environment. Linked to this international pressure grows to reduce demand on the earth's energy reserves. Modern technology has made great strides forward in developing new innovative heat sources but probably the greatest advance in combined thermal comfort

receding
逐渐远离

and energy conservation is the modern wet floor heating system. The advent of high quality plastics pipes has made possible the utilization of low temperature water in floor heating systems perfectly compatible with the new heat source technology. Fully developed for all types of floor construction, U.F.H. combines all types of heat, conduction, radiation and convection, matching the ideal temperature gradient throughout an entire building. The safe invisible, space saving vandal and tamper proof system is both responsive and energy conscious offering passive self regulation. Thermal comfort can be defined as the state of mind where satisfaction is felt with the thermal environment. Research shows that people feel most comfortable when their feet are a little warmer than their heads. Independent tests reveal that the most acceptable indoor climate is none in which the floor temperature ranges between 19~29 °C and the air temperature at head level ranges between 20~24 °C.

However, since individuality is integral to all human activity it is not possible to specify one set of environmental conditions which will meet all eases. The best results we are likely to achieve depend on a 5% dissatisfaction factor. There is no temperature that will please everyone, but we can aim to establish a comfort zone that will satisfy the highest possible percentage of those using this area.

With radiator or convector heating systems a vertical temperature gradient is produced colder at foot level than at the head. A modern indoor climate surely demands a heating system which will match the required conditions for human thermal comfort with the principal heating effect being evenly distributed at ground level and not above head level. We have seen that warm feet create good sensations, so let us examine the effect upon the indoor climate if we warm the whole floor to just the right temperature.

We have touched on improvements in building standards but no amount of insulation can change the laws of physics — heat still rises. Efficient insulation will, however, serve to trap heat above head level in an area where it can make no contribution to human comfort. Solving this problem involves a close study of the three types of heat available to us. Radiant heat provides the most pleasurable sensation of comfort. It contributes to the **exhilaration** of a walk in the spring sunshine even though the ambient air temperature may be only a few degrees above freezing. We humans also respond well to conducted heat — the cat-like pleasure that comes from the warmth of a hot water bottle or just **cuddling** up to another person. Lastly, there is convected heat caused by the effects of the radiation and conduction warming the air and causing it to rise. By using all three types of heat in association we can achieve very high levels

exhilaration
兴奋

cuddling
拥抱

of thermal comfort. The normal **criterion** for heating design is to achieve very high levels of thermal comfort. The normal criterion for heating design is to achieve a specified air temperature against the given heat loss of the building at a specified outside ambient temperature. When designing a floor heating system, however, lower air temperatures may be acceptable because of the higher level of overall radiation and the added benefit of conduction from warm, friendly floors.

In modern, well insulated buildings the temperature of the floor surface need be only just above air temperature in order to achieve required comfort factor. These low temperature differentials result in gentle, low velocity convection throughout the entire building. Low velocity convection reduces the amount of dust in the air in comparison with other types of heating. There are no inaccessible areas behind radiators or convectors where dust or dirt can collect. It is also cost efficient to operate. Eliminating high velocity convection means there will be no stack of high temperature air above head level.

A heated floor is a radiant plane: subjects standing on it will therefore receive the benefit of all round radiation.

A high level of radiant comfort means that air temperatures can actually be slightly lower with an infloor heating system than those usually

criterion
标准 准则

required for other methods of heating. Radiation, conduction and convection combine to create the ideal thermal environment for health and comfort.

本章重点内容介绍

现代人对人工环境的热舒适性要求很高。与此相关的是，国际上对地球能源储备需求的压力日益增大。现代技术在开发新的创新热源方面取得了长足的进步，但在热舒适性和节能性相结合方面最大的进步是现代湿地板供暖系统。

热舒适可以定义为对热环境感到满意的精神状态。研究表明，当人们的脚比头暖和一点时，他们感觉最舒服。测试表明，地板温度在 19~29 ℃，头部空气温度在 20~24 ℃的室内气温是最可接受的。没有任何温度能让每个人都满意，但我们可以建立一个舒适区，以满足使用该区域的人中可能达到的最高百分比。我们需要对三种可用热量进行仔细研究：辐射热提供最舒适的感觉。比如，在春天的阳光下散步，即使周围的空气温度可能只有零下几度，我们对传导的热量也有很好的反应。最后，由于辐射和传导的影响而产生的对流热使空气升温并使其上升。通过结合使用这三种热量，我们可以达到非常高的热舒适度。

热舒适性是在进行暖通空调设计时必须要达到的目标之一，随着现代人们对环境舒适性的要求变高，我们也应该更加注重满足不同人对热舒适的不同需求，通过对三种热量传递方式——辐射，对流，热传导的合理结合，去创造一个健康舒适的理想热环境。

Chapter 13
Hydronic Radiant Heating and Cooling
循环辐射供热和制冷

13.1 Hydronic Radiant Cooling (Commercial Buildings)
水力辐射制冷（商业建筑）

Cooling of non-residential buildings equipped with All-Air Systems significantly contributes to the electrical energy consumption and to the peak power demand. Part of the energy used to buildings is consumed by the fans that transport cool air through the ducts. This energy heats the conditioned air, and therefore adds to the internal thermal cooling peak load. Scientists at LBNL(Local Building National Laboratory) found that, in the case of the typical office building in Los Angeles, the external loads account for only 42% of the thermal cooling peak. At that time,28% of the internal gains were produced by lighting, 13% by air transport,

12% by people, and 5% by equipment. The implementation of better windows, together with higher plug loads due to increased use of electronic office equipment, have probably caused these contributions to change to some extent since then. HVAC systems are designed to maintain indoor air quality and provide thermal space conditioning. Traditionally, HVAC systems are designed as All-Air Systems, which means that air is used to perform both tasks. DOE-2 simulations for different California climates using the California Energy Commission (CEC) base case office building show that, at peak load, only 10% to 20% of the supply air is outside air. Only this small fraction of the supply air is in fact necessary to ventilate the building in order to maintain a high level of indoor air quality. For conventional HVAC systems the difference in volume between supply air and outside air is made up by **recirculated** air, as shown in Fig.13.1. The recirculated air is necessary in these systems to keep the temperature difference between supply air and room air in the comfort range. The additional amount of supply air, however, often causes draft as well as indoor air quality problems due to the distribution of pollutants throughout the building.

All-Air Systems achieve the task of cooling a building by convection only. An alternative is to provide the cooling through a combination of radiation and convection inside the building.

recirculated
再循环的

This strategy uses cool surfaces in a conditioned space to cool the air and the space enclosures. The systems based on this strategy are often called Radiative Cooling Systems, although only approximately 60% of the heat transfer is due to radiation. If the cooling of the surfaces is produced using water as transport medium, the resulting systems are called **Hydronic** Radiant Cooling Systems (HRC Systems). By providing cooling to the space surfaces rather than directly to the air, HRC Systems allow the separation of the tasks of ventilation and thermal space conditioning. While the primary air distribution is used to fulfill the ventilation requirements for a high level of indoor air quality, the secondary water distribution system provides thermal conditioning to the building. HRC Systems significantly reduce the amount of air transported through buildings, as the ventilation is provided by outside air systems without the recirculating air fraction. Due to the physical properties of water, HRC Systems remove a given amount of thermal energy and use less than 5% of the otherwise necessary fan energy. The separation of tasks not only improves comfort conditions, but increases indoor air quality and improves the control and zoning of the system as well. HRC Systems combine temperature control of the room surfaces with the use of central air handling systems.

hydronic
循环的

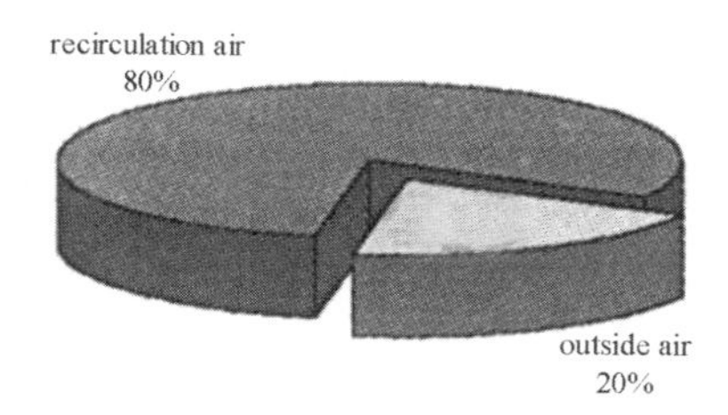

Fig. 13.1 Fraction of outside air and recirculation air for conventional all-air-system.

Due to the large surfaces available for heat exchange in HRC Systems (usually almost a whole ceiling, and sometimes whole **vertical** walls),the temperature of the coolant is only slightly lower than the room temperature. This small temperature difference allows the use of either heat pumps with very high coefficient of performance (COP) values, or of alternative cooling sources (e.g., indirect evaporative cooling), to further reduce the electric power requirements. HRC Systems also reduce problems caused by duct leakage, as the ventilation air flow is significantly reduced, and the air is only conditioned to meet room temperature conditions, rather than cooled to meet the necessary supply air temperature conditions. Furthermore, space needs for ventilation systems and their duct work are reduced to about 20% of the original space requirements. Beside the reduction of space requirements for the shafts that house the vertical air distribution system, floor-to-floor height can be reduced, which offsets the initial cost of the additional system.

vertical
垂直的

The thermal storage capacity of the coolant in HRC Systems helps to shift the peak cooling load to later hours. Because of the hydronic energy transport, this cooling system has the potential to interact with thermal energy storage systems (TES) and looped heat pump systems.

13.2 Hydronic Cooling Systems 水力冷却系统

Most of the HRC Systems belong to one of three different system designs. The most often used system is the panel system. This system is built from **aluminum** panels with metal tubes connected to the rear of the **panel**, as shown in Fig. 13.2.

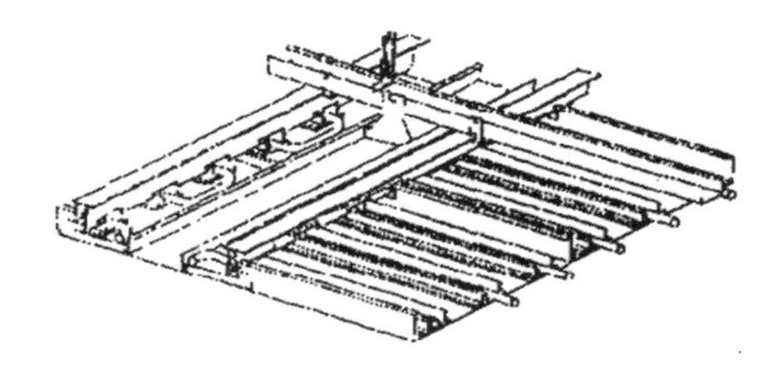

Fig. 13.2 Suspended panels figure courtesy of Flakt.

aluminum
铝

panel
嵌板，镶板

The connection between the panel and the tube is critical. Poor connections provide only limited heat exchange between the tube and the panel, which results in increased temperature differences between the panel surface and the cooling fluid.

Panels built in a "sandwich system" include the water flow paths between two aluminum panels (like the evaporator in a refrigerator). This arrangement reduces the heat transfer problem and increases the directly cooled panel surface. In the case of panels suspended below a concrete **slab**, approximately 93% of the cooling power is available to cool the room. The remaining 7% cools the floor of the morn above.

Cooling grids made of small plastic tubes placed close to each other can be imbedded in plaster, **gypsum** board, or mounted on ceiling panels (e. g. acoustic ceiling elements).

This second system provides an even surface temperature distribution. Due to the flexibility of the plastic tubes this system might be the best choice for retrofit applications. It was developed in Germany and has been on the market for several years, as shown in Fig. 13.3. When the tubes are imbedded in plaster, the heat transfer from above is higher than in the case of cooling panels.

The heat transfer to the concrete couples the cooling grid to the structural thermal storage of the slab. Plastic tubes mounted on suspended cooling panels show thermal performance comparable to the panel systems described above. Tubes imbedded in a gypsum board can be directly attached to a wooden ceiling structure

gypsum
石膏

slab
厚板，后块

without a concrete slab. Insulation must be applied to reduce cooling of the floor above.

A third system is based on the idea of a floor heating system. The tubes are imbedded in the core of a concrete ceiling, as shown in Fig. 13.4.

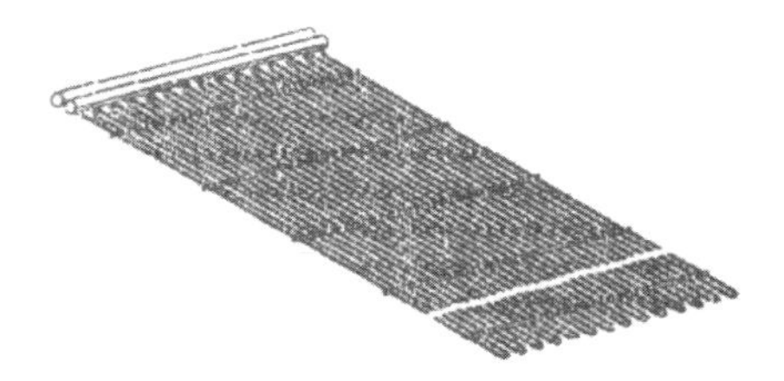

Fig. 13.3 Capillary tubes figure courtesy of KaRo information service.

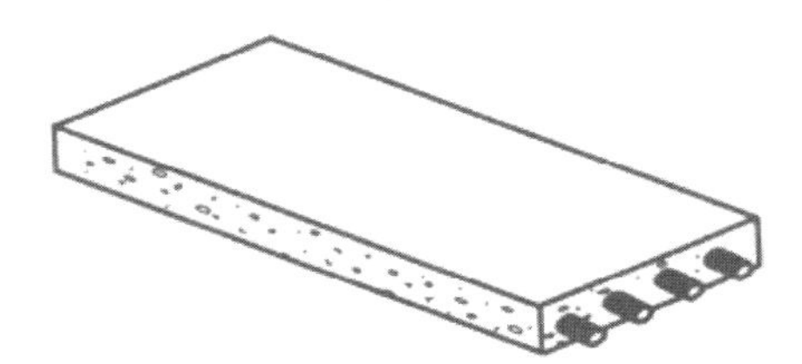

Fig. 13.4 Concrete core conditioning.

The thermal storage capacity of the ceiling allows for peak load shifting, which provides the opportunity to use this system in association with alternative cooling sources. Due to the thermal storage involved, the control of this system is limited. This leads to the requirement of relatively high surface temperatures to avoid uncomfortable conditions in the case of reduced cooling loads. The cooling power of the system is therefore limited. This system is particularly suited for alternative cooling sources, especially the heat exchange with cold night air. The faster warming of room with a particular high thermal load can be avoided by running the circulation pump for short times during the day to achieve a balance with rooms with a lower thermal load.

Due to the location of the cooling tubes in this system, a higher portion of the cooling is applied to the floor of the space above the slab. Approximately 83% of the heat is removed by the circulated water from the room below the slab, while 17% is from the room above.

Obviously, these three system types can also be used to heat a building.

13.3 Hydronic Radiant Heating and Cooling (Residential)

循环辐射加热和制冷（住宅）

While there are many examples of hydronic radiant heating and cooling installations in non-residential buildings available, very little has been reported about residential applications. For some time floor heating systems were installed in residences because of their superior thermal comfort. Besides thermal comfort issues, hydronic floor heating installations avoid recirculation air and the bothersome cycling of the air handling system. A positive side-effect is the elimination of duct leakage (which wastes up to 30% of conditioning energy) for naturally ventilated buildings or at least a significant reduction of the duct leakage effect for mechanically ventilated houses.

The installation of a floor heating system can also be used to cool the residence. There are several issues related to floor cooling:

(1) cooling power is limited due to small temperature difference between supply water and room air;

(2) floor surface temperature should not be less than 19 °C;

(3) the **dew point** of the indoor air has to be kept below the supply water temperature (condensation).

In humid climates, condensation can be avoided by dehumidifying the ventilation air either by mechanical cooling or by desiccant systems.

Floor heating and cooling are often designed as concrete core conditioning systems, which provide good heat transfer from the embedded tube to floor surface. Systems with the tubing installed between floor joists usually have a

dew point
露点

reduced heat transfer, which requires larger temperature differences between the supply water and the room air. While this might not be a problem when the system is in heating mode, large temperature differences in the cooling mode might create a danger for condensation or at least diminish energy savings because of the need to dehumidify the ventilation air to levels beyond thermal comfort criteria.

When designed well, hydronic radiant heating and cooling systems operate with temperatures close to design room air temperatures. When mated with a ground-source heat pump, these systems provide excellent energy efficiency. High heating supply water temperature and low cooling supply water temperature reduce the energy efficiency.

本章重点内容介绍

本章讲述了用于商业建筑的循环辐射冷却系统的定义、特点及应用，介绍了三种不同的循环冷却系统（面板系统，提供均匀的表面温度分布的系统，基于地板供暖系统的思想的系统）的原理。最后介绍了住宅的循环辐射供暖和制冷，循环地板供暖的优点，以及地板冷却存在的几个问题。

配备全空气系统的非住宅建筑的冷却显著增加了电能消耗和峰值电力需求。用于建筑的部分能量被通过管道输送冷空气的风扇消耗掉。这种能量加热空调空气，从而增加内部热冷却峰值负荷。暖通空调系统旨在保持室内空气质量并提供热空间调节。

所有的空气系统只能通过对流来冷却建筑物。另一种方法是通过建筑物内部的辐射和对流的结合来提供冷却。此策略使用空调空间中的冷表面来冷却空气和空间外壳。基于这种策略的系统通常被称为辐射冷却系统。如果表面的冷却是以水为输送介质，则产生的系统称为循环辐射冷却系统（HRC 系统）。循环辐射冷却系统降低电力需求，减少了管道泄漏引起的问题，通风系统及其管道系统的空间需求减少到原来空间需求的 20% 左右。除了减少垂直空气分配系统所用竖井的空间要求外，还可以降低楼层的高度，从而抵消附加系统的初始成本。HRC 系统中冷却剂的蓄热能力有助于将峰值冷却负荷转移到后期。由于循环能量的输送，这种冷却系统有可能与热能储存系统和循环热泵系统相互作用。

最常用的系统是面板系统。该系统由铝板制成，铝板后部连接有金属管。面板和管道之间的连接至关重要。连接不良只会限制管子和面板之间的热交换，从而导致面板表面和冷却液之间的温差增大。内置在“三明治系统”中的面板包括两个铝板（如冰箱中的蒸发器）之间的水流路径。这种布置减少了传热问题，增加了直接冷却的面板表面。对于悬挂在混凝土板下的面板，大约 93% 的冷却功率可用于冷却房间。剩下的 7% 冷却了上面的地板。

第二个系统提供均匀的表面温度分布。当管子嵌入石膏时，上面的传热比冷却板的传热要高。混凝土的传热将冷却网与板坯的结构蓄热耦合起来。安装在悬挂式冷却面板上的塑料管显示出与上述面板系统相当的热性能。嵌入石膏板中的管道可以直接连接到木质天花板结构上，而无需混凝土板。必须使用绝缘材料以减少上面地板的冷却。

第三个系统是基于地板供暖系统的思想。管道嵌入混凝土天花板的核心。天花板的蓄热能力允许峰值负荷转移，这为将该系统与其他冷却源结合使用提供了机会。要求相对较高的表面温度，以避免在冷却负荷减少的情况下出现不舒适的情况。因此，系统的冷却能力有限。该系统特别适用于替代冷源，特别是与夜间冷空气的热交换。通过在白天短时间运行循环泵，以实现与热负荷较低的房间的平衡，可以避免具有特定高热负荷的房间更快升温。由于冷却管在该系统中的位置，较高部分的冷却应用于板坯上方空间的地板。大约 83% 的热量通过循环水从板坯下面的房间排出，17% 来自上面的房间。

虽然在非住宅建筑中有许多可供使用的循环辐射供暖和制冷装置，但关于住宅应用的报道很少。除了热舒适性问题外，循环地板供暖装置还避免了空气循环和空气处理系统的设计。一个积极的副作用是消除自然通风建筑的管道泄漏（浪费多达 30% 的调节能量），或者至少显著降低机械通风房屋的管道泄漏效应。

Chapter 14
ASHRAE
美国暖通空调工程师协会

ASHRAE (American Society of Heating, Refrigerating and Air-conditioning Engineers) is the world's foremost technical society in the fields of heating, ventilating, air conditioning, and refrigerating. With more than 56,000 members from over 132 nations, ASHRAE is a diverse organization representing building system design and industrial processes professionals around the world.

ASHRAE advances the arts and sciences of heating, ventilation, air conditioning and refrigeration to serve humanity and promote a **sustainable** world. With more than 53,000 members from over 132 nations, ASHRAE is a diverse organization representing building system design and industrial processes professionals around the world.

sustainable
可持续的

ASHRAE's members are worldwide individuals who share ideas, identify needs, support research, and write the industry's standards for testing and practice. The result is that engineers are able to keep indoor environments safe and productive, while protecting and preserving the outdoors for generations to come.

One of the ways that ASHRAE supports the need of members and industries for information is through ASHRAE Research. Thousands of individuals and companies support ASHRAE Research annually, enabling ASHRAE to report new data about material properties and building physics and to promote the application of innovative technologies.

14.1 Introduction
简 介

ASHRAE, founded in 1894, is a global society advancing human well-being through sustainable technology for the built environment. The Society and its members focus on building systems, energy efficiency, indoor air quality, refrigeration and sustainability within the industry. Through research, standards writing, publishing and continuing education, ASHRAE **shapes** tomorrow's built environment today. ASHRAE was formed as the American Society of Heating, Refrigerating and Air-Conditioning Engineers by the merger in 1959 of American Society of Heating and Air-Conditioning Engineers (ASHAE) founded in 1894 and The American

shape
计划

Society of Refrigerating Engineers (ASRE) founded in 1904.

14.2 ASHRAE's Mission and Vision

美国暖通空调工程师协会的使命和愿景

(1) Mission

To advance the arts and sciences of heating, ventilation, air conditioning and refrigeration to serve humanity and promote a sustainable world.

(2) Vision

ASHRAE will be the global leader, the foremost source of technical and educational information, and the primary provider of opportunity for professional growth in the arts and sciences of heating, ventilating, air conditioning and refrigerating.

14.3 ASHRAE' s Core Values

美国暖通空调工程师协会的核心价值观

(1) Excellence

ASHRAE education, technical information and all other activities and products will always reflect the best practices that lead our industry. We strive for continuous improvement and innovation in all our practices and products.

(2) Commitment

ASHRAE and its members are passionate about serving the built environment, creating value, and recognizing the accomplishments of others.

(3) Integrity

ASHRAE is committed to the highest ethical standards. We work transparently, observing essential requirements for due process and peer reviews to assure our members and stakeholders that we do the right things the right way.

(4) Collaboration

ASHRAE seeks and embraces collaborative efforts with organizations, agencies, and individuals sharing our commitment to sustainable built environments.

(5) Volunteerism

Members lead ASHRAE at every level, serving ASHRAE and helping ASHRAE serve society.

14.4 Professional Development and Certification
专业发展与认证

More than 2,500 **certifications** have been

certification
证明、证书

earned by professionals who have demonstrated their knowledge and expertise in the HVAC&R industry. Gain a competitive edge by earning one of the following ASHRAE certification types:

14.4.1 Building Commissioning Professional Certification 建筑调试专业认证

Building Commissioning Professional Certification (BCxP) validates competency to lead, plan, coordinate and manage a commissioning team to implement commissioning processes in new and existing buildings.

More than 2,750 certifications have been earned by professionals who have demonstrated their knowledge and expertise in the HVAC&R industry. Gain a competitive edge by earning an ASHRAE certification in building commissioning.

The BCxP certification, an ANSI-Accredited Personnel Certification Program under ISO/IEC 17024 (#1139), validates competency to do the followings:

Lead, plan, coordinate and manage a commissioning team to implement commissioning processes in new and existing buildings.

The BCxP certification has been recognized

by the U.S. Department of Energy (DOE) as meeting the Better Buildings Workforce Guidelines(BBWG).

(1) ASHRAE's BCxP certification program validates competency against **scheme** requirements for the Building Commissioning Professional set forth in the Better Buildings Workforce Guidelines.

(2) Elevate your reputation among peers, in the workplace and among clients.

(3) Demonstrate competency in critical building commissioning knowledge, skills and abilities.

To ensure the continued validity and relevance of its Commissioning certification scheme, beginning with the 2017 class of renewals, ASHRAE's CPMP program will validate continuing competency against scheme requirements for the Building Commissioning Professional set forth in the Better Buildings Workforce Guidelines. CPMPs who renew against these scheme requirements will earn the Building Commissioning Professional (BCxP) designation.

For CPMPs, the benefits of **recertifying** as a BCxP versus applying for the BCxP certification are a shorter exam and a lower application fee.

scheme
计划

recertifying
矫正

14.4.2 Building Energy Assessment Professional Certification 建筑节能评估专业认证

Building Energy Assessment Professional Certification (BEAP) validates competency to assess building systems and site conditions; analyze and evaluate equipment and energy usage; and recommend strategies to **optimize** building resource utilization.

More than 2,750 certifications have been earned by professionals who have demonstrated their knowledge and expertise in the HVAC&R industry. Gain a competitive edge by earning an ASHRAE certification as a building energy auditor.

The BEAP certification, an ANSI-Accredited Personnel Certification Program under ISO/IEC 17024 (#1139),validates competency to assess building systems and site conditions; analyze and evaluate equipment and energy usage; and recommend strategies to optimize building resource utilization.

The BEAP certification has been recognized by the U.S. Department of Energy (DOE) as meeting the Better Buildings Workforce Guidelines (BBWG) for the Commercial Building Energy Auditor.

optimize
使……最优化

(1) Validate competency against scheme requirements for the Building Energy Auditor certification set forth in the Better Buildings Workforce Guidelines.

(2) Meet Federal Buildings Personnel Training Act (FBPTA) requirements in up to 50 competency area performances, as identified by the General Services Administration (GSA).

(3) Demonstrate competency in critical commercial building energy auditor knowledge, skills and abilities.

(4) Elevate your reputation among peers, in the workplace and among clients.

(5) Comply with local, state and federal requirements.

(6) Conduct Building EQ in Operation ratings.

14.4.3 Building Energy Modeling Professional Certification 建筑节能建模专业认证

Building Energy Modeling Professional Certification (BEMP) validates competency to model new and existing building and systems with the full range of physics; and evaluate, select, use, calibrate and interpret the results of energy modeling software where applied to building and systems energy performance and economics.

More than 3,000 certifications have been earned by professionals who have demonstrated their knowledge and expertise in the HVAC&R industry. Gain a competitive edge by earning an ASHRAE certification in building energy modeling.

Developed with the participation of the U.S. **affiliate** of the International Building Performance Simulation Association (IBPSA-USA) and the Illuminating Engineering Society (IES), the BEMP certification, an ANSI-Accredited Personnel Certification Program under ISO/IEC 17024 — Accreditation # 1139, validates competency to do the following:

(1) Model new and existing building and systems with the full range of physics.

(2) Evaluate, select, use, calibrate and interpret the results of energy modeling software where applied to building and systems energy performance and economics.

① Demonstrate competency in critical energy modeling knowledge, skills and abilities.

② Elevate your reputation among peers, in the workplace and among clients.

affiliate
隶属于

③ Comply with local, state and federal requirements.

④ Conduct Building EQ as Designed ratings.

14.4.4 Certified HVAC Designer 暖通空调认证设计师

The Certified HVAC Designer (CHD) certification validates competency to do the following: Design HVAC systems to meet building/project requirements, including equipment, equipment sizing, load calculations, mechanical equipment room design, duct and piping design and layout, and develop HVAC plans for permit and construction.

More than 3,000 certifications have been earned by professionals who have demonstrated their knowledge and expertise in the AC&R industry. Gain an ASHRAE competitive edge by earning an ASHRAE certification in HVAC design.

The Certified HVAC Designer(CHD) certification validates competency of the HVAC Certified Designer, working under the responsible charge of an engineer, to do the following:

Design HVAC systems to meet building/project requirements, including load calculations, equipment selection and sizing, mechanical equipment room design, duct and piping design and layout for the development of HVAC plans for permit and construction.

Why is earning the ASHRAE Certified HVAC Designer (CHD) credential the best way to gain a competitive edge?

In a recent "Industry Need" survey, ASHRAE Member respondents who influence the HVAC Designer hiring decision had this to say:

(1) HVAC Designer certification is a tool to identify competent new hire prospects (74%).

(2) It is a worthwhile professional development goal (82%).

HVAC Designers themselves agree that earning an HVAC Designer certification would be a "worthwhile professional development goal" (75%) and "help differentiate practitioners from their peers" (70%).

Earning the Certified HVAC Designer (CHD) certification will let your employer, peers and customers know that you have the knowledge, skills and abilities needed to get the job done, and position you for continued recognition and success.

14.4.5 High-Performance Building Design Professional Certification 高性能建筑设计专业认证

High-Performance Building Design Professional Certification (HBDP) validates competency to design and integrate sustainable HVAC&R systems into high performing buildings.

More than 3,000 certifications have been earned by professionals who have demonstrated their knowledge and expertise in the HVAC & R industry Gain an ASHRAE competitive edge by earning an ASHRAE certification in high-performance building design developed with the participation of the Illuminating Engineering Society (IES) and the Mechanical Contractors Association of ANSI America(MCAA) and with input from the Green Building ACCREDIIED.

Initiative(GBD) and the U.S. Green Building Council (USGBC), the HBDP certification, an ANSI — Accredited Personnel Certification Program under ISO/EC 17024 — Accreditation #1139, validates competency to do the following:

Design and integrate sustainable HVAC&R systems into high performing buildings.

(1) Demonstrate competency in critical high-performance building design knowledge, skills

and abilities.

(2) Elevate your reputation among peers, in the workplace and among clients.

(3) Comply with local, state and federal requirements.

14.4.6 Healthcare Facility Design Professional Certification 医疗保健设施设计专业认证

Healthcare Facility Design Professional Certification (HFDP) validates competency to incorporate standards, guidelines and **regulatory** codes as well as unique healthcare facility requirements and design principles in HVAC system design.

More than 3,000 certifications have been earned by professionals who have demonstrated their knowledge and expertise in the HVAC&R industry Gain an ASHRAE competitive edge by earning an ASHRAE certification in healthcare facility design developed with the participation of the American Society for Healthcare Engineering Certified (ASHE) of the American Hospital Association, the HFDP certification validates competency to do the following:

regulatory
调整的、监管的

Incorporate standards, guidelines and

regulatory codes as well as unique healthcare facility requirements and design principles in HVAC system design.

(1) Demonstrate competency in critical healthcare facility design knowledge, skills and abilities.

(2) Elevate your reputation among peers, in the workplace and among clients.

14.4.7 Operations & Performance Management Professional Certification 运营与绩效管理专业认证

Operations & Performance Management Professional Certification (OPMP) validates competency to manage facility operations and maintenance to achieve building performance goals, including those related to indoor environmental quality, health and safety.

More than 3,000 certifications have been earned by professionals who have demonstrated their knowledge and expertise in the HVAC&R industry. Gain a competitive edge by earning an ASHRAE certification in operations and performance management.

Developed with the participation of APPA and

the General Services Administration (GSA), the OPMP certification validates competency to do the following:

Manage facility operations and maintenance to achieve building performance goals, including those related to indoor environmental quality, health and safety.

(1) Meet Federal Buildings Personnel Training Act (FBPTA) requirements in up to 98 competency area performances, as identified by the General Services Administration (GSA).

(2) Demonstrate competency in critical building operations and performance management knowledge, skills and abilities.

(3) Elevate your reputation among peers, in the workplace and among clients.

美国供暖和空调工程师协会(ASHAE)于1894年成立，美国制冷工程师协会(ASRE)于1904年成立。美国供暖和空调工程师协会（ASHAE）、美国制冷工程师协会（ASRE）是前身，后来发展、整合，最终形成今天的美国暖通空调工程师协会（ASHRAE）。

美国暖通空调工程师协会是一个多元化的协会，会员遍布世界各地，是世界上在供暖、通风、空调、制冷等领域领先的技术协会。ASHRAE拥有来自132个国家的56 000多名成员，是一个多元化的组织，代表着世界各地的建筑系统设计和工业过程专业人员。ASHRAE 推动供暖、通风、空调和制冷的艺术和科学，为人类服务，促进形成一个可持续的世界。

ASHRAE 的成员分享想法，提供需求，支持研究，并为测试和实践编写行业标准。工程师们致力于保证室内环境的安全和生产，是技术和教育信息的首要来源，促进暖气、通风、空调和制冷等学科的科学发展。

References
参考文献

[1] Chung O, Jeong S G, Yu S,et al. Thermal performance of organic PCMs/micronized silica composite for latent heat thermal energy storage[J]. Energy and Buildings, 2014, 70: 180-185.

[2] Feldman D, Banu D, Hawes D,et al. Obtaining an energy storing building material by direct incorporation of an organic phase change material in gypsum wallboard[J]. Solar Energy Material, 1999, 22(2): 231-242.

[3]Hawes D W. Latent heat storage in concrete, PhD Thesis. Concordia University. Montreal, Quebec, Canada, 1991.

[4] Hawes D W, Feldman D. Absorption of phase change materials in concrete[J]. Solar Energy Material and Solar Cell, 1992, 27(2): 91-101.

[5] Sari A, Karaipekli A, Kaygusuz K. Capric acid and stearic acid mixture impregnated with gypsum wallboard for low-temperature latent heat thermal energy storage[J]. International Journal of Energy Research, 2008, 32: 154-60.

[6] Shi X, Memon S A, Tang W C, et al. Experimental assessment of position of macro encapsulated phase change material in concrete walls on indoor temperatures and humidity levels[J]. Energy and Buildings, 2014, 71: 80-87.

[7] Sandra C, José A, Victor F, et al. Mortars based in different binders with incorporation of phase-change materials: Physical and mechanical properties[J]. European Journal of Environmental and Civil Engineering, 2015, 19(10): 1216-1233.

[8] Haurie L, Mazo J, Delgado M,et al. Fire behavior of a mortar with different mass fractions of phase change material for use in radiant floor system[J]. Energy and Buildings, 2014, 84: 86-93.

[9] Cengel Y A, Boles M A. Thermodynamics An Engineering Approach [M]. 8th ed.McGraw Hill Education, 2014.

[10]Van Ness H C. Understanding Thermodynamics [M].Dover Publications, 1983.

[11] Holman J P. Heat Transfer (McGraw-Hill Series in Mechanical Engineering) [M]. 10th ed. McGraw Hill Education, 2009.

[12] Bergman T L, Lavine A S, Incropera F P, et al. Introduction to Heat Transfer [M]. 6th ed. John Wiley and Sons Ltd, 2011.

[13] Frank P. Incropera, David P. DeWitt, Theodore L. Bergman, Adrienne S. Lavine. Fundamentals of Heat and Mass Transfer [M]. 6th ed.John Wiley and Sons Ltd, 2006.

[14] Yunus A. Cengel, John M. Cimbala. Fluid Mechanics: Fundamentals and Applications [M]. 4th ed. McGraw Hill Education, 2017.

[15] Frank M. White. Fluid Mechanics[M]. 8th ed. McGraw-Hill Education, 2015.

[16] 沈维道，童钧耕 . 工程热力学 [M].5 版 . 北京：高等教育出版社，2016.

[17] 杨世铭，陶文铨 . 传热学 [M]. 4 版 . 北京：高等教育出版社，2007.

[18] 吴望一 . 流体力学 [M]. 北京：北京大学出版社，1982.

[19] 王革，薛若军，顾璇 . 建筑环境与设备工程专业英语 [M]. 哈尔滨：哈尔滨工程大学出版社，2007.

[20] 张喜明，王浩，赵嵩颖，于文艳，安玉华 . 建筑环境与能源应用工程专业英语 [M]. 2 版 . 北京：中国电力出版社，2018.

[21] 白雪莲，刘猛，李文杰 . 建筑环境与能源应用工程专业英语 [M]. 重庆：重庆大学出版社，2014.

[22] 王方，张仙平 . 建筑环境与能源应用工程专业英语 [M]. 北京：中国建筑工业出版社，2019.

[23] 圆山重直 . 传热学 [M]. 王世学，张信荣，等，译 . 北京：北京大学出版社，2011.

[24] 圆山重直 . 工程热力学 [M]. 张信荣，王世学，等，译 . 北京：北京大学出版社，2011.